The Guerrilla's Guide to the Baofeng Radio

NC Scout

Radio Recon Group

For SFC Samuel Hairston and my late Father.

Table of Contents

1. Introduction to the Baofeng Radio

The Baofeng radio, in all of its incarnations, is an inexpensive, simple to use, Very High Frequency (VHF) / Ultra High Frequency (UHF), Frequency Modulation (FM) low-power analog radio that can provide an incredible capability with minimal training time.

The task and purpose of this manual is to provide the reader with a comprehensive guide not just operating the radio but to the broader topic of improvised line of sight (LOS) communications in the field in a potentially non-permissive, high threat environment. The fact that the Baofeng, a the time of this writing, may be the most prolific handheld radio in the world cannot be ignored; the guerrilla, the dissident, the political activist, must arm himself with what he has rather than what he wished he had. This reality is frequently referred to as the Common / Off The Shelf (COTS) approach to warfare. In most cases, the Baofeng in one of its most common forms will be the lone communications option and the potential fighters must know how to best employ it. This manual addresses it and more importantly, applying the communications techniques taught to Special Operations Forces to its use. A radio must be considered as valuable a tool as any other weapon.

This manual is written from and for the perspective of anyone to use, regardless of skill level, implementing sound communications techniques absent more sophisticated infrastructure. The techniques are written from an end-user perspective with technical data kept to a minimum and only emphasized where absolutely necessary. It is explicitly written with the potential guerrilla in mind. This is not a book on Amateur Radio, however, much can be carried over to the

Amateur Radio (Ham) hobby as a means to supplement and practice many of the topics of instruction contained. Some of the techniques described are for academic or information purposes only, until they are required to keep a people free. The author takes no responsibility in the manner in which the techniques are used.

The Three Roles of Communications:

Before diving into communications as a topic it is critically important to understand the purpose behind the need to communicate. It has been my experience that this is frequently lost on most either just starting out or with no frame of reference. There are three overlapping yet distinct roles of communications, each with their own unique considerations:

- **Sustainment**
- **Tactical**
- **Clandestine / Strategic**

Sustainment Communications are communications we create in the absence of conventional phones. This role includes public safety. This is what is being referenced when the topic of survival communications is being discussed, whether that's for establishing a localized network on a rural retreat, small communities or in the wake of a natural disaster. Sustainment communications seek the maximum range of coverage possible between groups due to the unpredictable nature of day-to-day life, emulating or replacing the convenience of a cell phone. This should under no circumstances be confused with Tactical communications. Communications security (COMSEC) is relatively low, although not completely ignored. The same tactics, techniques and procedures (TTPs) discussed in this book are applicable to Sustainment-level Communications.

Tactical Communications are those used when conducting military-type operations and require immediate response, ie coordinating fire between two teams on a target, conducting friendly linkup in the field, and relaying short-range battlefield information to higher echelons. Tactical Communications are immediate in nature and require a far higher level of communications security, relying on brevity or transmission time, the use of codes, and higher-loss antennas to shorten the range as much as possible in order to limit who may be hearing the transmissions beyond the intended recipients.

Clandestine or Strategic Communications require the highest degree of planning and coordination as well as the highest level of communications security. These are normally transmitted over a much longer range and considerations to prevent interception are at a maximum.

Clandestine communications are those coordinating the activities of an underground group operating in resistance to a governing entity with a superior level of resources. This category includes otherwise overt transmissions with ulterior motives, coded transmissions sent in a short burst of data, and highly directional antennas to further mitigate the potential for interception. This manual will go into these methods and how to implement them with a Baofeng radio in explicit detail.

Baofeng Radio Capabilities:

The Baofeng radio in most configurations operates in three frequency ranges with a fourth limited to a few specific models:

- 67-108 megahertz (mHz) RECEIVE ONLY
- 136-174 mHz TRANSMIT / RECEIVE
- 400-470 mHz TRANSMIT / RECEIVE
- 220-250 mHz TRANSMIT / RECEIVE ***Certain Models***

67-108 mHz is the international broadcast FM radio range. Inside the US the broadcast FM radio range specifically is 87-108mHz, with the frequency of each station ending in an odd decimal. This is obviously useful for monitoring local commercial radio but has other purposes as well.

First, an underground can set up a clandestine low power FM radio transmitter, set to an unused frequency, to create a local pirate radio station for resistance information and friendly propaganda. If those are set to an even decimal, only those equipped with Baofeng radios will clearly hear it (*since conventional FM broadcast radio is in odd decimals*). Once a resistance cell creates and distributes a net schedule, a local pirate radio station can quickly get on, and off, the air.

Second, the frequency range for military ground communications is:

- **30-88 mHz US / NATO**
- **30-108 mHz CSTO / RUSSIA / China / Venezuela**

Since the first half of the FM broadcast monitoring range of the Baofeng is also the last half of both NATO and CSTO standard ground communications systems, the Baofeng in each of its incarnations can be used to intercept or monitor these transmissions.

An important note - just because a transmission is encrypted does not mean it cannot be intercepted. The presence of traffic, even if only unintelligible digital squelch, in an area of operations can be an early warning. Insurgents in Iraq frequently used analog TVs operating in the 40-55mHz range with the volume turned off, simply watching for changes in static as an early warning for the presence of American troops. The same can be done with the Baofeng.

Frequency Considerations

The Baofeng operates as a dual band radio. This means is transmits in both VHF and UHF in a fairly wide range of frequencies ranging from those requiring various types of licenses to those that do not. Each come with their own set of issues, ranging from over-reliance on published data to volunteering too much sensitive information. The answer lay with not establishing a consistent pattern while recognizing the full range of the equipment's capability. Inside the frequency ranges the radio transmits are various services that will contain different types information and unique traffic.

Very High Frequency

136-174 mHz is the Very High Frequency (VHF) range of the Baofeng radio. VHF performs well in rural, hilly environments and is less susceptible to signal loss in dense vegetation. It also has a higher degree of reflectivity off of major terrain features. VHF typically has a longer range than the higher frequency Ultra High Frequency (UHF) and has a higher power output, regardless of power setting, from the transmitter. The VHF frequency range is broken down further into specific services:

- Multi-Use Radio Service (MURS)
- Business Band VHF
- NOAA Weather
- Marine Band
- 2 Meter Amateur Radio

Multi-Use Radio Service (MURS) is a license-free set of five specific frequencies (channels). Since it is license free, MURS is favorable to low-key, low-cost sustainment level communications for those not requiring a high level of sophistication. The MURS frequencies are:

- **MURS 1 :** 151.820
- **MURS 2 :** 151.880
- **MURS 3 :** 151.940
- **MURS 4 :** 154.570
- **MURS 5 :** 154.600

Because MURS is both openly published and channelized, it is not always recommended that they be used alone in an unconventional warfare (UW) environment. That said, it can offer strong capability with minimal training required.

Business Band VHF is a licensed service using specific frequencies for commercial enterprises for transmitting both voice and data. They are channelized and the frequency range is:

- 151.505 MHz

- 151.5125 MHz
- 151.625 MHz
- 151.700 MHz
- 151.760 MHz
- 151.955 MHz
- 154.515 MHz
- 154.540 MHz
- 158.400 MHz
- 158.4075 MHz

It is critical to recognize these may be shared between businesses and that emergency service communications are also in between these frequency ranges. Therefore it is critical to monitor these ranges prior to making any communications plans.

NOAA, or the National Oceanic and Atmospheric Administration, utilizes seven frequencies that transmits weather data (and sometimes emergency relief information) for its given region twenty-four hours a day. The seven frequencies are:

- 162.400 MHz
- 162.425 MHz
- 162.450 MHz
- 162.475 MHz
- 162.500 MHz
- 162.525 MHz
- 162.550 MHz

Marine Band radio, as the name implies, is designated for use among maritime vessels. It utilizes the frequency range of 155-

160mHz, and channelized for specific roles. Marine Band, when inland, is often used by public service agencies including law enforcement.

High Band VHF Amateur Radio, also known as **2 Meter**, is nicknamed after the physical length of one full wavelength of the frequency when transmitting. In the amateur (*or Ham*) radio world this is how specific bands are named and makes antenna calculations simple (*more on this in the antenna chapter*). Amateur radio requires a license to operate and utilizes the 144-148 mHz frequency range.

A word about Cover For Action (CFA)

While Amateur Radio requires a license to operate lawfully, and creates a significant amount of personal data behind a license, it also creates a useful cover for action (CFA) for otherwise suspicious activities guised as having a benign purpose.

Nellie Ohr obtained a Technician-level Ham radio license (KM4UDZ) before becoming involved in the leaking of a dossier involving then-President Donald Trump. It has been alleged that in northern VA and DC, where a high amount of radio monitoring equipment exists for counterintelligence purposes, that she and her husband were instructed by British freelance spy Christopher Steele to obtain a Ham radio license in order to transmit the files without having to do it through otherwise monitored or potentially compromised means. Were they caught, the cover for action would have been simple:

We were obtaining equipment for a fun hobby and the testing of ham radio equipment on a digital mode.

Covers for action are just as dependent on the investigator not knowing enough about the topic as it is the plausibility of the cover.

Ultra High Frequency

400-470 mHz is the Ultra High Frequency (UHF) range of the Baofeng radio. UHF performs well in urban environments. In rural environments UHF experiences higher signal loss in dense vegetation. It also has a tendency to penetrate buildings rather than reflect and refract signals off of them (discussed further in the operating techniques section). That said, for those reasons UHF can be very effective at mitigating interception when used at the tactical level.

An enemy with signals intelligence (SigInt) interception and exploitation equipment have a more difficult time first intercepting then gaining a line of bearing (LOB) on a low power UHF signal in dense vegetation - but that said, a tactical group must also understand their physical range in terms of distance between teams must also be shorter and often less than 1k meters when also using low power. UHF is broken down into the following services:

- Family Radio Service
- General Mobile Radio Service
- UHF Business Band
- 70cm Amateur Radio

Family Radio Service (FRS) and **General Mobile Radio Service** (GMRS) operate within the same frequency space. These are the frequencies the common two way radios found in sporting good stores use. The former does not require a license, but has a low power output (2 watts). GMRS requires a license that covers a group, allowing repeater use and up to 50 watts of power output. The frequency range for FRS / GMRS:

- **1:** 462.5625
- **2:** 462.5875
- **3:** 462.6125
- **4:** 462.6375
- **5:** 462.6625
- **6:** 462.6875
- **7:** 462.7125
- **8:** 467.5625
- **9:** 467.5875
- **10:** 467.6125
- **11:** 467.6375
- **12:** 467.6625
- **13:** 467.6875
- **14:** 467.7125
- **15:** 462.5500
- **16:** 462.5750
- **17:** 462.6000
- **18:** 462.6250
- **19:** 462.6500
- **20:** 462.6750
- **21:** 462.7000
- **22:** 462.7250

Because FRS / GMRS is channelized with known published data, it should be used with the same considerations to COMSEC as MURS. Like MURS, however, it provides a solid option for simple sustainment level capability with minimal training.

UHF Business Band is the counterpart to the VHF version. The frequencies are:

- 464.500 MHz
- 464.550 MHz
- 467.850 MHz
- 467.875 MHz
- 467.900 MHz
- 467.925 MHz
- 469.500 MHz
- 469.550 MHz

70cm Amateur Radio, like its two meter VHF counterpart, is in reference to one full wavelength of a UHF signal in that portion of the band. Its frequency range is 420-470 mHz and is incredibly popular in most of the more densely populated areas of the US and the world.

220-250 mHz

220-250 mHz, also known as the 1.25m Band when referencing the Amateur Radio portion (220-225 mHz), is featured on certain models of Baofeng radios called ‘Tri-Band’. The frequencies themselves are situated in the high end of the VHF band, getting close to the beginning of the UHF range. It offers excellent performance in terms of range and clarity in woodland environments superior to UHF but not quite as well as the lower VHF band.

This capability is important to note however due to the fact that few radios operate in this particular frequency range. Because there is a shortage of equipment, little attention is paid to this portion of the spectrum. For this reason alone it can mitigate both the likelihood of interception as well as exploitation. I will emphasize, however, that while it offers a level of security through obscurity, this should never mean basic practices of COMSEC should be overlooked.

Summary

The Baofeng radio offers a vast amount of versatility in an inexpensive package, enabling much capability in the hands of small groups with the skill to utilize it in the proper way. This manual seeks to identify how to use the radio in the field with minimal infrastructure, creating a competent communications plan, communications techniques, a guidebook on improvised antennas, and how to implement digital communications and encryption for the maximum communications security capability. This is not and should not be considered a book on Ham radio. It is the application of communications for guerrilla operations.

2. Functions and Field Programming

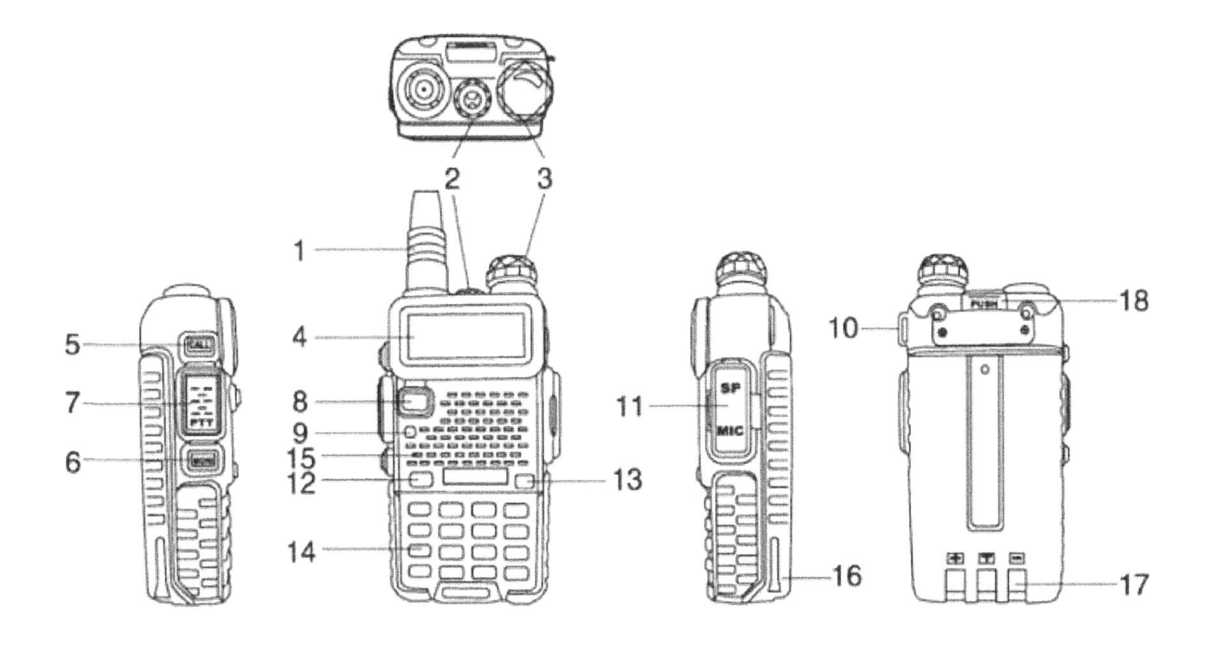

1. antenna
2. flashlight
3. knob (ON/OFF,volume)
4. LCD
5. SK-side key1/CALL(radio,alarm)
6. SK-side key2/MONI(flashlight,monitor)
7. PTT key(push-to-talk)
8. VFO/MR (frequency mode/channel mode)
9. LED indicator
10. strap buckle
11. accessory jack
12. A/B key(frequency display switches)
13. BAND key(band switches)
14. keypad
15. SP.&MIC.
16. battery pack
17. battery contacts
18. battery remove button

Basic Controls

UV-5R's layout is simple - an on / off knob that controls volume, a key pad, a push to talk (PTT) and a couple other buttons with unique functions on the side. Every aspect of the radio is controllable via these buttons and the embedded menus.

Because every aspect of the radio is controllable and configurable absent software, it is very simple to troubleshoot in the field once the operator becomes experienced with the basic controls. Each button has a maximum of only two functions, making it simple for corrective action.

Frequency and Memory Modes

Right after unboxing, the first control to pay attention to is the VFO / MR button located just below the screen. VFO stands for *Variable Frequency Oscillator*, and is a throwback term to the old days of radio when everything was crystal controlled - a crystal oscillated at a certain frequency and that was the one the radio was operating on. VFO simply is another term for frequency mode, and the MR, also known as Memory or Channel mode, is the list of frequencies programmed into the radio's memory banks. When in Frequency mode you can directly enter the frequency you want to operate on, in Channel mode you're selecting which channel you've per-programmed into the memory.

When the UV-5R is in channel mode there will be small numbers to the far right on the display. These are the programmed channels based on their position on the list. When in frequency mode, those numbers are not present, however, if the radio is using a narrowband frequency, it will have two very small numbers just to the right of the large digits representing the frequency. On that note, often newer folks to radio confuse the terms 'channel' and 'frequency' - know that these are not interchangeable. One is a specific operating frequency, the other is a frequency entered into a memory or a frequency assigned to a specific band or radio service. It might sound complicated, but its not.

A short word on Communications Security, Counter-Intelligence and Memory Modes

All of the data you input into these radios are potentially exploitable. What that means, without mincing words, is that if captured will absolutely be used against you. Every adversary on the potential battlefield should be considered an intelligence collector, formal training or not, and your radio is one of the most valuable targets carried.

With that said, every Taliban member with a radio in Afghanistan became a target of interest for us. Not just because of the presence of the radio, but the manner in which they were communicating. They were primarily using analog dual band radios very similar to the Baofeng. Our practice for recovering a radio was to not turn it off and immediately hand it off to the signals intelligence collector we had assigned to the team who'd then break down all of the contents in the memory in an effort to map their current communications plan.

Do not, under any circumstances, program the memory of a radio you intend on using for tactical or clandestine purposes. These will be used against you when compromised, in addition to leading to complacency in practice among your group. I can tell you from experience that complacency does not just kill you, but kills all of the ground gained by your movement. From a counterintelligence perspective, we want to give a potential adversary as little as possible, and at every opportunity make them believe we are in a position of weakness and inferior capability.

For this reason I strongly advise groups to input their operating frequencies in the Frequency mode alone and do not bother with the memory at all. If everything is a factory default, should the radio be captured it will provide almost no value to the group that captured it. This same logic applies to only using license-free, channelized data as well. It creates patterns, and the root of intelligence analysis is pattern recognition. While programs like Chirp make streamlining communications planning seem easy, it also becomes a serious crutch leading to training scars in practice. In my experience, in addition to the exploitation of data, it also leads to team members not knowing how to troubleshoot their own equipment and usually inducing failures when trying to do so. The Baofeng is a simple piece of gear. Keep it that way.

Baofeng Front Panel Programming

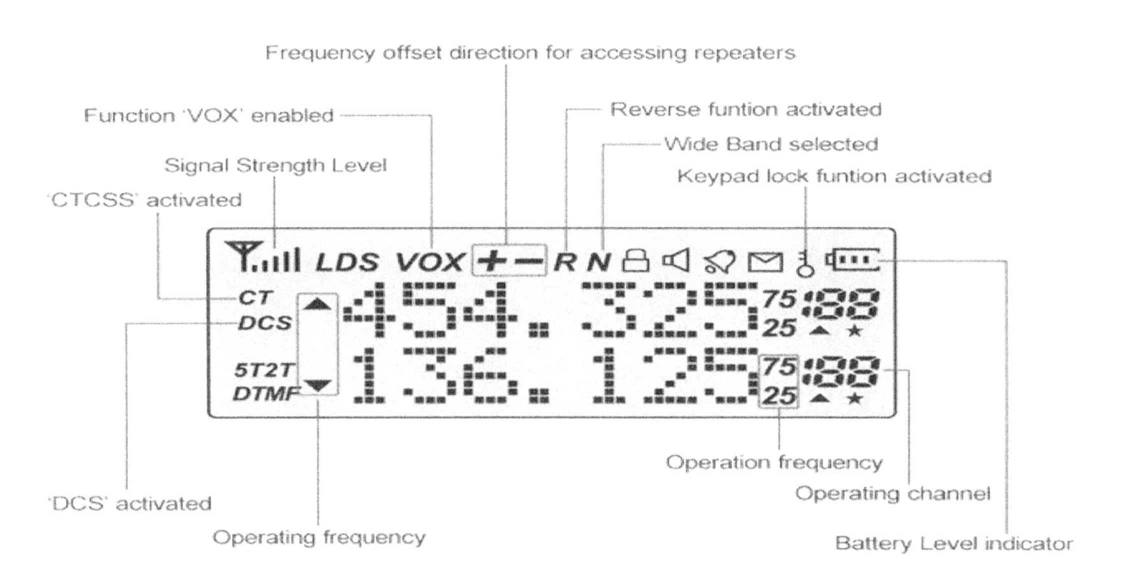

The screen is a simple LCD with a top and bottom frequency, a battery life indicator, and a small signal strength meter in the top corner. The other symbols are various indicators for menu functions enabled, such as VOX, power level, and repeater offsets. The arrows to the left of the frequency numbers indicates which frequency you're operating on (top or bottom), unless its blinking - then it indicates the radio is receiving on that particular frequency.

When in Channel Mode the radio also has the option of displaying the name of the channel in alpha-numeric letters. People frequently do this for naming the license-free frequencies, repeaters for Amateur Radio or GMRS operations, or particular frequencies in a dedicated communications plan.

Entering or Changing a Frequency

To get going in a hurry, keep the UV-5R in frequency mode by hitting that VFO/MR button. If the voice prompt is enabled in the menus (*more on this later*), it will tell you Channel Mode or Frequency Mode. From here, you can directly enter the frequency on which you're wanting to transmit or receive. Simply punch in the numbers and you're good to go - but with that said it requires six digits, or it won't complete the action. For example, a frequency of 144.100 must be entered with all six digits to change. If you mess up, no worries, either press the EXIT button or just let it set for a couple seconds, it'll go back to the old frequency and you can start the process over.

If you are in Channel Mode, the digits you press correspond to the channel assignment in the memory list. Looking for channel 12? Enter 1 then 2 on the keypad. Looking for channel 2? Enter 0 then 2. That simple.

Hit the A/B button (number 12 in the diagram above) to select between the top and bottom frequencies or channels. You'll know which one you're on by the black arrow out to the left of the numbers.

Again, if you get hung up, just give the radio a second, and it will return to the old operating frequency. It's really that simple, but I often have people who get frustrated when entering frequencies into the radio for the first time because they're overthinking the equipment. In my experience its not the radio that gets messed up, its a lack of knowledge on the part of the operator.

Lock That Keypad!

The Radio Operator must be able to troubleshoot his gear, especially at night. Further, he must also have the experience level to identify the vectors of issues before they arise - just as with weapons training, we have to know what causes the malfunctions in an effort to prevent them before it costs us. In one small unit tactics course I was teaching some time back, the class was conducting a movement to contact; ie, a patrolling technique that is searching for an unknown enemy along a planned route. The movement was taking place at night. Like all patrols as we pointed out in Chapter 1, they were split into two groups of roughly 10 guerrillas each; an assault element and a support by fire. Having several well seasoned combat vets from the Army and the USMC in the class, their movement was predictably picture perfect. Quiet, precise, and aware. They moved like stone cold killers, and I was proud to observe it.

Patrol Leader's radio then emitted a sound; a short beep, just loud enough for me to hear. Despite his experience level as a recently retired Infantry Platoon Sergeant, with these types of patrols being almost reflexive from his twenty years of doing it, he was wholly unfamiliar with his UV-5R. He didn't lock the control panel before movement, and it shifted frequency when bounced around on his gear during movement. It happens. With comms now lost at night, the two elements of the patrol began to increase the distance between one another. Anything beyond 100m in poor visibility or at night is a serious liability. He called a halt and was visibly frustrated at losing command and control. Having even more of a spread between the groups, he attempted to walk to the trailing support by fire group - and predictably, they shot him, thinking he was an enemy. Welcome to the reality of night time patrolling.

Now this was in training, and he was shot with blanks. But the lessons learned were very much real. In training, you'll die a lot. Without communications, you'll die, a lot. And one of the skills that needs to be hammered home is the ability to program and troubleshoot your equipment on the fly. Expect no one to be able to do it for you. In an unsupported operational environment, be it the mostly benign rural retreat in the first scenario or a nigh time ambush, or anything in between, you have to have the working knowledge for every in and out of the radio itself just as you would your weapon system or any other piece of equipment you're carrying. Because your life depends on it.

What I always advocate doing, and might be a great habit to get into, is locking the keypad after each change I make. Just as in the example I shared, your radio absolutely will change frequency settings on you while you're moving. This is really regardless of radio; once, when I was a newly minted Infantry Fire Team Leader, we were experimenting with EF Johnson public safety UHF radios for inter-team use. We did an air assault insertion on a training exercise and I didn't have comms with anyone. Those EF Johnsons did not have a front panel but just a knob for changing the pre-programmed channels, and mine was one off a couple channels. My M4 hit it when I was exiting the Blackhawk. Things happen, mistakes are made and lessons are learned. I learned a lesson, and like my Patrol Leader in the previous example, I learned to lock down all the functions of my radio before movement.

On a Baofeng we do this by holding down the # key, which will give you a voice prompt saying LOCK (if enabled) and there will be a small key icon that appears beside the battery indicator. Now you'll notice that all of the keypad functions have been disabled aside from the push-to-talk (PTT). To disable the LOCK, hold down the # key again until it says unlock and the icon will disappear.

It is important to note that the lock function does not control volume. It can still get knocked off. Another student and friend of mine who's an active duty Marine Communications NCO in an ANGLICO unit, which are high speed Marines responsible for calling in artillery and air munitions, was in my Scout Course leading another night patrol. They lost comms. But this time the culprit wasn't an unlocked screen, but the volume knob getting cut all the way down without him knowing it. He learned a valuable lesson about gear that night and so did everyone else.

Other functions of note:

The top side key (#5 in the diagram above), just above the Push To Talk (PTT), has two functions. A quick press switches it to receive broadcast FM. Holding the button down enables an alarm, which transmits the same alarm tone on the frequency the radio is set. This feature is useful for a quick alert inside a group on that net or for exploiting (active jamming) potential enemy groups using the same frequencies. It can accidentally be triggered by inexperienced users, and a second long press of the button shuts it off.

The button underneath the PTT (button #6) also has two functions. A short press cuts on the most annoying feature on the radio - the tiny LED light on top. Two quick presses makes it a strobe. There’s a number of ways to physically disable this, but I simply paint it red and cover it with duck tape. The color red is the shortest wavelength of light and the duck tape keeps it from emitting much light.

A long press of this button activates the Monitor, which is just the opening of the squelch. This is used for the reception of weak signals. If a signal is coming in broken, you can open the squelch with the monitor and hear everything coming over that frequency, no matter how weak or down in the static it may be.

Antenna Port and Microphone Jack

The Baofeng in all of its forms utilizes an SMA-Male antenna connection point. These are small with fine threads and while functional, they are far from rugged. The stock antenna is adequate for communications on both bands within a mile or so, but are frequently replaced with longer antennas to increase range in terms of distance. This is fine for sustainment communications; for tactical communications extending the physical range may be a bad thing in certain cases. This is covered in depth in the antenna and operations chapters. More importantly, the physical strength and dexterity of the antenna connector itself becomes a problem when used with longer antennas.

I replace all antenna connections with BNC adapters. BNC adapters feature twin binding posts and are far more rugged in the field when using both extended antennas and purpose built wire antennas with coax cable. Further, BNC adapters have a larger surface area at the base of the connection, relieving any stress that may be put on the antenna connection inside the radio. Threading is no longer required, it is a simple twist on and twist off.

The microphone jack is located on the side of the radio opposite the PTT. It is better known as a Kenwood two prong connector and a large aftermarket exists for microphones. This is also the data port for software programming the radio via the programming cable and the jack used for data bursts via the K1 cable covered in Chapter 7.

Quick Programming Menu Options

The Baofeng is simple in operation and has a unique advantage in that every function of the radio is controllable via either the front panel itself or the menu. There are 40 total menu options. Most of them are irrelevant for the actual functionality of the radio.

All menu items are selectable by first hitting the MENU button then using the UP and DOWN arrows to cycle through them. You can also hit the number of the menu item to go directly to it. By pressing MENU again, you can change the options of that specific menu number.

Once you've selected which option you want, hit MENU again to save it.

Menu 0: Squelch

Squelch is simply a filter that keeps the radio from being keyed up from weak signals. Since we want to hear the most number of signals while not listening to static, keep the squelch set at 1.

Menu 1: Step

Step, or frequency step, the spacing between specific frequencies. This is more important to the scanning feature. I suggest setting it to 6.25kHz.

Menu 2: TXP (Transmit Power)

Transmit power allows you to select which power setting the radio will transmit on. Low emits 1 watt, high is 4 watts, 8 watts or 10 watts, depending on the model. Some of the higher powered Baofengs, such as the 8HP and the AR-152, have a medium and high setting. The general rule of thumb is for sustainment communications the highest power setting is good to go, for tactical and clandestine purposes the low power setting is more favorable to prevent interception.

Menu 4: VOX

VOX, or voice activation, enables sound picked up by the microphone to key the radio rather than the PTT. This is useful for digital operations covered in Chapter 6. For all purposes other than digital operation, set VOX to off.

Menu 7: TDR

Transmit / Dual Receive (TDR) enables the radio to receive on both displayed frequencies at once. This is a requirement for split mode operation; ie, receiving on one frequency and transmitting on another as a simple means of communication security.

Menu 8: BEEP

This controls the beeping noise made by the key pad. I turn these off.

Menu 9: TOT

TOT, or Time Out Timer, shuts the radio off after an extended time of transmitting. I set this to 60, or 60 seconds. This prevents extended times that operators may hotmic, or accidentally trigger their radios.

Menu 10-13: DCS / CTCSS

These control the sub-audible tones emited and received by the radio. CTCSS in particular can be used as a simple means to prevent communications jamming but is easily picked up by communications receivers and frequency counters. The most frequent use for this in the United States is for Amateur Radio repeater operation, which are marked as PL tones. These are set to prevent accidental activation of the repeater where many of the same uplink and downlink frequencies overlap one another in a given area. I set them all to OFF to avoid confusion in the field.

Menu 14: Voice

Voice is simply the voice prompt when cycling through the menus or entering frequencies. The three options are ENG (English), CHI (Chinese) or OFF. Select OFF.

Menu 23: BCL

BCL, or Backlight, controls the backlighting of the screen. For tactical use in the field, shut this off.

Menu 25: SFT-D

This controls the standard offset for repeater functions. If you’re not using repeaters, also known as simplex operation and what the bulk of this manual covers, set this to OFF.

Menu 26: Offset

This controls the offset frequency itself. The standard amateur radio frequency offsets for VHF and UHF repeaters are 000.600 (600kHz) and 005.000 (5mHz), respectively. If you're setting up your own repeater system, as covered in Chapter 6, you can set your own offset known only to your group.

Menu 29-31: LED Controls

These control the backlight being triggered by either transmitting or receiving. The settings are BLUE, ORANGE, PURPLE, and OFF. As with Menu 23, I set these to OFF.

Menu 39: ROGER

This function controls what is, at least in my opinion, the most annoying feature of any radio: the roger beep. Its the tone that gets transmitted after someone is done transmitting. It should be set to OFF on all radios.

Menu 40: Reset

This is one of the most important features of the radio - the factory reset. This is similar to the Z function on military radios - deletes all settings in the memory and anything that may have been saved, preventing any potential exploitation. Sometimes a factory reset is necessary to correct any issues with the radio. These are rare but do happen. Its default setting is ALL or just the VFO. When resetting, reset ALL. When you do this, you'll have to manually reset each of the settings.

Summary

With the menu functions themselves now covered, troubleshooting the radio on the fly should become simple in the field with training and practice. Keep it simple - the radio is a simple piece of gear that requires consistent training to stay sharp as with any other critical task. With that said, should problems arise, make sure each of the menu items are set correctly as noted in this checklist. It will save you headaches down the road.

3. Communications Planning

Proper communications technique first relies on the ability to plan. We call this a Signals Operating Instructions (SOI). The SOI is a type of chart that logically lays out a group's communications in an easy to follow way between teams operating in a given area and the headquarters, known as a Tactical Operations Center (TOC) coordinating their activities.

The SOI includes multiple layers - the frequency tables themselves, the callsigns used by the individual teams, passwords and a simple encryption grid for authentication of transmissions known as a **Search And Rescue Numerical Encryption Grid (SARNEG)**. Each of these layers add higher degrees of communications security (COMSEC). One of the biggest failures of the Global War On Terror was mostly abandoning these methods with the excuses of digital encryption and expediency in lieu of proper tradecraft and sound techniques. Both have led to a severe complacency in the face of a possibly technologically superior adversary. If it is your turn to be the Guerrilla, the entire way of thinking must return to the practices perfected during the Cold War.

The SOI is, along with the planned route, the most sensitive part of a group's operations plan. It is trashed after each mission - an SOI is not to be reused. Units tasked with Unconventional Warfare in Vietnam had multiple sets of frequencies per mission in recognition that the Vietcong and NVA had significant levels of technology dedicated to intercepting them. The SF Patrolling Tips Guide, B-52, suggested certain practices for teams operating for extended periods in the jungle and those suggestions have been refined into the sample SOI flow chart below.

The SOI includes the PACE plan of frequency sets, each labeled with their own codeword. The Primary and Alternate frequency sets have a transmitting (TX) and receiving (RX) frequency. On the Baofeng this is simply done by ensuring the radio is first in Frequency Mode then Transmit / Dual Receive (TDR / Menu #7) is enabled. The transmitting frequency is entered in the top, the receive frequency is entered on the bottom. This is reversed for the other group - and creates a frequency set. When operating this way an adversary must identify both frequencies in the set in order to intercept the entire transmission. While not impossible by any means, this adds a layer of protection. The Mode is also identified in case a method other than voice, such as a digital mode for a data burst, is being used. If it is simply voice, label it as voice.

The second half of the SOI, the callsigns, passwords, SARNEG and Commo Windows provide instructions on how and when to communicate both over the radio and in person. Its a simple, logical way for a team to map out capabilities and is a critical part of the mission planning process. At the conclusion of a guerrilla operation the entire plan is scrubbed, never to be used again.

Intelligence collectors and analysts tasked with counterguerrilla operations put an emphasis pattern recognition and recovering an SOI is considered a jackpot. For that reason the SOI also laid out in the manner it is - so that even in the presence of more sophisticated equipment and capabilities, the underlying tradecraft in the techniques are sound, making life difficult for an enemy. Do not fall into the trap of thinking because one plan worked at a given point of time it will then become the default plan for all future operations. Complacency kills. While sustainment communications considerations may get away with this, tactical and clandestine level communications absolutely cannot.

Sample SOI Chart:

Signals Operating Instructions (SOI)

DURATION:

Primary [FREQ / MODE]:

CODEWORD:

TX:

RX:

Alt. Freq [FREQ / MODE]:

CODEWORD:

TX:

RX:

Contingency Freq:

Emergency Signal:

Callsigns:

TOC / Control:

Element 1:

Element 2:

Support/Recovery:

Challenge / Password:

Running Password [DURESS]:

Number Combination:

Search and Rescue Numerical Encryption Grid [SARNEG]:

K	I	N	G	F	A	T	H	E	R
0	1	2	3	4	5	6	7	8	9

***** SARNEG first number rotates with last numeral of previous date*****

COMMO WINDOW SCHEDULE:

	1	2	3	4	5	6	7	8	9	10

- **DAY:**
- **NIGHT:**

PACE Planning

PACE stands for **Primary**, **Alternate**, **Contingency** and **Emergency** and is a four layered approach to every aspect of planning a mission, from route selection, support equipment, to communications. Having four options in the field are critical not just from a capabilities approach but also for communications security and a team's survival during personnel recovery (PR) phase of an operation. The role of each layer of a PACE plan:

- **PRIMARY:** Primary method of communications
- **ALTERNATE:** Used if Primary is no longer functional
- **CONTINGENCY:** Initiates Personnel Recovery Plan
- **EMERGENCY:** Non Electronic / Near and Far Recognition for Linkup

Primary and **Alternate** are essentially interchangeable. Both of these have the transmit and receive frequencies listed along with the mode of operation. If its voice, it will be marked voice, for digital, identify which digital protocol (ex. MT-63, DMR, THOR, etc.). There could be a wide variety of reasons for communications failure during a mission that are completely outside a team's control; Out of range, changing environmental conditions, or electronic warfare by a sophisticated enemy. This is resolved by switching to a new frequency or set. Primary and Alternate frequency sets are assigned a code word; teams indicate switching between the frequency sets by simply saying the code word to initiate the changeover.

Example:

Alternate frequency set is assigned with Codeword: ALICE

"Wolf, this is Hound, Over"

"Hound, this is Wolf"

"ALICE"

"ALICE, Roger"

groups enter new frequency set

In the event both primary and alternate frequency sets fail, the mission objective changes. History being a consistent guide, a failure in communications will lead to a team getting killed. The mission then changes to Extraction and Personnel Recovery (PR).

Contingency is the frequency set aside for PR and is assigned a unique codeword so that minimal instructions are required over the net. Anyone familiar with civil aviation instantly knows MAYDAY, which is probably the best known example. Said three times on a specific frequency (121.5 mHz), it immediately triggers a recovery / rescue effort. This is labeled as the Running Password or Duress word in the SOI and works the exact same way.

When planning missions in Iraq and Afghanistan, teams would always have a Contingency frequency which was also labeled the Medical Evacuation (MEDEVAC) frequency, which was a single channel rather than frequency hopping and unencrypted. We did this for absolute reliability when lives are on the line - for that reason, the Contingency line of an SOI must also be a single frequency. Traffic on this frequency is kept to a minimum as well - the team is under fire or has been compromised, we will be linking up at the pre-planned site for our extraction. All of this is triggered with the pre-planned codeword. Planning, again, is critical.

Emergency is always a non-electronic signal that is used to signal a team's location to the group coming to rescue them. It includes a Day and Night signal. There's a large number of ways to do this but the two most common tactical level signals are using a VS-17 (blaze orange / hot pink flag) during the day and a colored chem lights at night. The signal is used in two ways - marking the team's location and sending a signal that they are not being held hostage.

To make linkup, the team that sees their signal will give a responding signal. This is known as Near and Far Recognition Signals. The teams then link up, exchange bona fides (the challenge and password from the SOI) and return to friendly lines.

Callsigns

Callsigns are a word, numbers, letters, or some combination thereof to identify the origin of traffic on a network. On the civilian side, callsigns are used to identify an individual station through its government assigned letters and numbers. They are issued to individuals, clubs and special events for Amateur Radio, to groups with GMRS, and businesses or organizations for business band use. There is much personal data attached, including the license holder, the geolocation of the registration, and in the case of amateur radio, the region assigned to the callsign.

In the conventional side of the military, callsigns are used to indicate the organization and the authority level in that organization's hierarchy. For example, when I was assigned to a Light Infantry Company early in my career, all of the callsigns were APACHE, designating Alpha Company. There was no other APACHE company in the Battalion or Brigade. APACHE 6 was the Company Commander, 5 was the Executive Officer (second in command), 7 the First Sergeant (1SG), and 1-4 the individual Platoon Leaders. APACHE 1-7 would be the Platoon Sergeant, 1-1 the First Squad Leader, etc. These remained static and did not change, indicating the position of the person communicating on the net.

Unconventional Warfare (UW) and Guerrilla Operations are a completely different story. Intelligence collectors and analysts map organizations based on hierarchy. For that reason every effort must be made to conceal the levels of authority and the actual roles within the organization when communicating.

Callsigns, when used for communications in a Guerrilla War, must frequently be changed with each SOI. One technique I recommend is having the callsigns follow a theme, each related to one another while leaving little indication of their actual nature or position of authority. The names themselves can be literally anything, but callsigns must be assigned to, at a minimum, three distinct roles:

1. The Tactical Operations Center (TOC), operating on the authority of the Ground Force Commander;

2. The Maneuver Elements (Teams) conducting the operation;
3. The dedicated Personnel Recovery (PR) and Support Element.

Again, the words used as callsigns themselves are irrelevant. Understanding the role for each, covered in the communications briefing during the planning phase of an operation, is critical, as is making every effort to conceal the role or purpose of the transmission to a potential signals intelligence collector.

Passwords, SARNEG and Authentication

Challenge and the accompanying password are used in person only; they are never said over the radio. These serve the purpose of identifying friendlies when linking up in the field. Even if a group knows one another and is very familiar with the group they're joining, a challenge and password becomes a critical way to signal if a linkup is dangerous or not. The way these are passed is one person approaching another from the two groups and stating the challenge, followed by the other person stating the password.

In WWII this was the predominant method used by the French and Danish resistance forces against the Nazi occupation and have been used by countless underground and clandestine groups every since. It is a simple way to give one group a heads up whether it is safe to make a linkup, or not.

We call these exchanging of bonafides; using the words in short phrases, which, like the callsigns, are meaningless by themselves. If the challenge and password is dog and horse, for example, they would be exchange the words as:

"Excuse me, did you see the lost dog poster?"

"No, I thought they were looking for a horse."

Again, these are not to be used over the radio. It is a method for linking up in person only. Should the wrong password be given twice, no further attempt to make a linkup should be made and that group should be considered compromised.

Running Password

The second type of password is the Running Password, or Duress word, to indicate a team in distress. When this is used the TOC or Commander immediately switches to the Contingency frequency to coordinate the recovery of the team. This is commonly done in both civil aviation and maritime traffic for aircraft or vessels in distress using the word MAYDAY three times over the air. Most people are familiar with this method and it works the same way for small groups, albeit with a different word.

SARNEG

The SARNEG, is the proper method for authenticating traffic over the air. Each letter being assigned a number, and the sequence of those numbers rotating daily, allows for a simple and robust method of authentication over the air. Due to the multiple ways it can be used and the fact that it changes every 24 hours, it can be used over and over for a several day duration of a mission.

Given the sample SOI SARNEG above, KINGFATHER, we would contact the receiving station and initiate the contact with their callsign followed by the authentication code:

“Hound, this is Wolf, over”

“Wolf, this is Hound”

“Authenticate KILO, ONE, over”

“I authenticate ZERO, INDIA, over.”

Since Kilo (K) is assigned a Zero and One is assigned India (I), the station answering responds with the corresponding numbers and letters (said phonetically) over the air. As with the password, if this is incorrectly given twice, the responding station could be indicating that they have been compromised without blatantly giving it away.

The SARNEG also has the purpose of encrypting number values during a message. Grids to a location are the most frequent use but this also applies to numbers of items as well. A grid square in latitude / longitude, Military Grid Reference System (MGRS) or Universal Transverse Mercator (UTM)can use this method to replace the numbers with letters which can only be decoded in that given 24 hour time period with the current SARNEG. For example, a MGRS grid when sent in a report would look like:

MD 1552 8719

MD IAAN EHIR

Any information gathered from this is essentially useless when taking into account that first, it will change in 24 hrs, and second, the enemy would have to have prior knowledge of what is even being transmitted. Since there’s a significant difference in each of the grid locator methods, if those are even used, the information is mostly useless by the time it is processed by a signals intelligence unit.

Communications Windows

During an operation, the two most dangerous times for a group on the ground is during the overt phases of movement - insertion into the operating area and during extraction. While on the move, all radio equipment that a team has on the Patrol is switched on and is being actively monitored in case of enemy contact. Once they move into their planned hide site, patrol base or safe house, a report is sent up indicating their safe arrival and that they will be checking in at the next pre-planned commo window. Radios are switched off to preserve battery life and mitigate the possibility of accidental transmissions from the hide. And the team will only communicate via radio from an area outside of their hide site, known as a transmission site. This further obscures their presence and the techniques for this is covered in Chapter 5.

Communications Windows, or commo windows, are pre-planned periods of time that teams will check in with the TOC or with other teams in the field. They are typically once every 12 or 24 hours, depending on the mission requirements, and are usually two hour blocks of time. The actual transmission can be sent anytime during that two hour block to further randomize the amount of time on the air.

In Afghanistan we would regularly have two commo windows in a 24 hour period; one for daylight, one for night. If two were missed this would automatically signal something wrong with the team on the ground and trigger the personnel recovery plan. A good friend of mine missed two in a row during a mission and this was extremely out of character - we knew something had happened to him and his team. We approached his team's last known location (their hide site overwatching a Taliban-aligned village on the border with Pakistan) and their team, despite their high level of training and discipline, had fell victim to the cold and the creeping sleep that happens when you get warm in a cave. They slept through the first commo window, and then did the same for the second. At least they were alive.

Communications Planning Considerations

The methods described here are robust in detail and assume a high threat environment. While not necessary in this detail for sustainment level communications, in a tactical or clandestine environment this degree of planning will become absolutely necessary regardless of the level of sophistication capability of an enemy. It requires discipline; failure to recognize this reality will result in compromise and eventual failure. This is a reminder also that the methods for planning an SOI ensure that a group does not fall into the trap of creating a pattern; changing ones practices regularly stymies and frustrates the efforts of predictive analysis hunting you, and in doing so, will keep you alive.

4. Traffic Handling and Reports

There's a right way and a wrong way to communicate on the air. Despite anything else, the biggest rule of thumb is to keep it short and sweet- the shorter the amount of time on the air, the better. Further, the clearer and more precise the language and annunciation on the air, the easier it will be for the receiver on the other end to understand you. We do not know what we sound like on the other end, and in addition, we have to assume two distinct but competing, concepts:

1. Everything we are saying is being monitored; and,
2. Everything we're saying is being copied down.

There must be some reason we're transmitting. There is no such thing as routine traffic - if you're breaking squelch in an unconventional warfare environment, you're running the risk of being intercepted and found. While this may not be important for sustainment level communications, it is absolutely vital at the tactical and clandestine operations level. The faster and clearer we can convey the information the more effective we become, and this begins with understanding how to send traffic the correct way.

"You, this is Me, Over"

At the beginning of each transmission the radio operator is calling another station to make contact - the operator first says the callsign of the station he's calling then his own. You, this is me,

Over. The responding station will come back with the calling station's callsign first, then theirs, as an acknowledgment.

"Wolf, this is Hound, Over"
"Hound, this is Wolf, go ahead, over"

Prowords

Prowords are a series of words used in a transmission with a specific meaning. They are meant to keep the conversation short with as few explanations or unnecessary chatter to a minimum. The most common prowords and their corresponding meanings:

OVER:	**I am done with this statement, over to you**
OUT:	**I am done transmitting**
ROGER:	**I understand**
COPY / HOW COPY:	**I copied this down / What did you copy**
BREAK:	**I am going to the next line / I am breaking in**
SAY AGAIN:	**Say what you said again**
WILCO:	**I will comply**
AUTHENTICATE:	**Give me the corresponding SARNEG combination**
REPEAT:	**Send the same firing solution (field artillery)**

The use of prowords keeps things short and precise. That said, they also indicate a level training and organization. So while in a military sense, old habits die hard (and for good reason), they can also tip off anyone tasked with interception and monitoring. The words themselves only have meaning because they are in common practice; for a potential guerrilla, understand the need for them and if substitutes can be created, use them, if needed. Just a consideration.

As a historical example of this concept, Russian ground commanders had a difficult time collecting anything of value from their signals intelligence assets when targeting Chechen leadership during the first and second Chechen wars. The Chechen leadership cadre in the first war was primarily made of veterans of the Soviet - Afghan war and had prior knowledge of not just the Russian capabilities but their failures in Afghanistan. They primarily leaned on local dialect to relay their traffic and supplied a large number of commercial VHF radios to all of the villagers in targeted areas, creating a large volume of false or useless traffic and thus concealing the important information.

Recognizing their own training and organizational needs, while also understanding its pitfalls, allowed them a strong degree of initial success. In addition local slang reinforced the morale of the people, reminding them they were in a struggle for survival. Both factors are critical notes for successful Guerrilla leadership.

Phonetics

As with prowords, the use of phonetics becomes critical on the air. We do not know what we sound like on the other end; that said, D and E sound the same when we say them over the air (DEEE, EEEE, etc.) so to avoid confusion phonetics are used in place of letters and numbers.

NATO Standard Phonetic Alphabet:

Letter	Word	Spoken	Letter	Word	Spoken
A	Alfa	Al fah	N	November	No vem ber
B	Bravo	Brah voh	O	Oscar	Oss cah
C	Charlie	Char lee	P	Papa	Pah pah
D	Delta	Dell tah	Q	Quebec	Keh beck
E	Echo	Eck oh	R	Romeo	Row me oh
F	Foxtrot	Foks trot	S	Sierra	See air rah
G	Golf	Golf	T	Tango	Tang go
H	Hotel	Ho tell	U	Uniform	You nee form
I	India	In dee ah	V	Victor	Vik tah
J	Juliett	Jew lee ett	W	Whiskey	Wiss key
K	Kilo	Key loh	X	Xray	Ecks ray
L	Lima	Lee mah	Y	Yankee	Yang key
M	Mike	Mike	Z	Zulu	Zoo loo

Number	Spoken
0	Zee roh
1	Wun
2	Too
3	Tree
4	Foh wer
5	Fife
6	Six
7	Sev en
8	Ait
9	Nin er

The military world uses the NATO standard, and commonly in law enforcement other words, usually names, are used in place of the letters themselves.

Using a standardized set of phonetics keeps the transmission time short by cutting down any potential confusion on the air. While using the NATO standard can give away a certain degree of training and organization to a signals intelligence collection and analysis team, this potential risk does not outweigh the benefits. Since the potential guerrilla’s best asset is unpredictability, creating your own set of standard phonetics inside a group may be ideal. Just keep in mind, this can create problems when coordinating with other guerrilla bands across a region. As long as there is one set standard everyone adheres to, a group will be good to go.

Reports

Going back to our first concept above - that what we are transmitting is being written down - owes to the necessity of transmitting mission critical information. The primary purpose of this is to transmit intelligence in the form of the location and capability of an enemy force. We do this through a simple report format known as SALUTE and its supplemental report, SALT.

SALUTE:

S: Size in number of people (PAX) and Vehicles (VICS)

A: Activity / What are they doing?

L: Location of activity

U: Uniform / how can we positively identify (PID) them?

T: Time of observation

E: Equipment / Details on weapons, equipment, and capability

SALT

S: Size / PAX & VICS

A: Activity

L: Location

T: Time of observed activity

Everyone with a radio, regardless of purpose, should be at least familiar with the format of these reports and the method to send them. The more eyes, the better. SALUTE reports are a simple way to relay that information in a clear, digestible format that allows an analyst to paint a more accurate picture in terms of an enemy's capability.

At first glance, SALUTE and SALT seem the same; there is much overlap between the two. SALUTE reports are the first and most detailed report to be sent, and create the baseline. The SALT report supplements the SALUTE report for any changes that may have occurred while observing the target.

Sample SALUTE Report:

"Wolf, this is Hound, Over"

"Hound, this is Wolf. Authenticate Kilo One, Over"

"Wolf, this is Hound. Zero India, SALUTE Report, Over"

"Roger, Standing by"

"Line Sierra: 7 PAX, 1 VIC, Break"

"Line Alpha: Conducting Pre-Movement Checks, Break"

"Line Lima: Mike Delta India Alpha Alpha November, Echo Hotel India Romeo, Break"

"Line Uniform: Brown hats, Black hoodies, red arm band, Break"

"Line Tango: One Two Five Seven, Local, Break"

"Line Echo: Plate carriers, AR-15s with IR Laser, Helmets with NOD mounts. How Copy, Over?"

The team reported what they observed in detail, line by line. At the end of each line they utilized Break to indicate that they were done and going to the next line, while keeping their physical descriptions short and precise. Also used was the SARNEG for the location as well as indicating the time zone (LOCAL) for the time of observation.

Going back to the primary purpose of SALUTE and SALT reports, understand that the handing off of this information is intended first to give an intelligence analyst the tools he needs to make an accurate assessment of the capabilities of the observed enemy, or, an assessment on whether they're an enemy at all. For the group on the ground, the primary question becomes how much information do we need to positively identify (PID) a target when handing them off to other groups.

In Afghanistan we used SALUTE and SALT extensively to hand off targets between our teams operating in a given area. When we conducted patterns of life missions against suspected Taliban villages (watching a target, collecting all pertinent information to collect a timeline of behaviors before attacking them) we would routinely write out SALUTE reports specifically naming identifying characteristics of individuals of interest. This way, what one team could see (and the others could not), we were able to hand off to maintain eyes on target. This is a primary training method in surveillance and reconnaissance.

Once, while observing a target, we had one team observe two men emplacing an IED on a route into the target village - I could not see them from my position, and they were outside engagement range. All they could do was mark the site of the IED and call it up in a SALUTE report along with PID on the two targeted individuals. Once they completed laying the IED and moved back to the village, we engaged them and the other armed Taliban that came to their aid from the village. All of this was possible with an accurate, disciplined SALUTE report.

All of this underscores the need for accurate reporting. Since the bread and butter of the guerrilla lay with the concept of mobile warfare, ie streamlining the roles of reconnaissance and combat as seamlessly as possible while conserving as much of your own resources as possible, the SALUTE report plays an instrumental role in identifying targets of interest and conversely, the ones to bypass. It goes without saying attacking hardened troops, such as line Infantry units of an occupation force, is a particularly bad idea especially early in the guerrilla war phases. Support troops, however, present a far more lucrative and usually successful target, with the added psychological benefit of sapping the morale of a larger force dependent upon them. Ergo, a guerrilla's training must heavily rely upon observation and discernment.

Medical Evacuation Report

One of the unfortunate realities of combat is the casualties it produces. Alongside that reality is that for a guerrilla band, you must take every effort to first recover your own and have a follow on method of treating them. While the continuum of care under fire to the guerrilla hospital are well outside the scope of this manual, the handling of Medical Evacuation (MEDEVAC) requests are not.

For the best reference on the care under fire to the guerrilla hospital continuum, I direct readers to COL Rocky Farr's *"Death of the Golden Hour and the Return of the Guerrilla Hospital"* first and to *"Where There Is No Doctor"* second, while also stating bluntly that a guerrilla movement, to succeed, requires first basic and competent medical treatment training for its combatants and second the recruitment of Doctors for the welfare of the people within its areas of interest.

Che Guevara was a Doctor by training in Argentina. Based on his own work in "The Motorcycle Diaries", despite his upbringing in an upper class Left leaning household in Argentina, his true radicalization into Marxist revolution came about during his traveling residency administering medical aid to the poor. It would become a staple of his plan to foment revolution in Cuba, where it was most successful, but also in the Congo and later in Bolivia. Despite his role as a guerrilla force commander, his administering of medical aid to the impoverished through free clinics fomented favorability among the target populace. It has been a model used by Leftists and adopted by the US as doctrine both in Special Forces and in Civil Affairs since. It works.

With that said the treatment of battlefield casualties must be at the center of a guerrilla commander's planning should he wish to be successful in maintaining morale. Fighters have reason to believe in a cause, especially one in which they are fighting for justice and social change, when they see the efforts a commander makes to recover and care for them under duress. Thus, the MEDEVAC becomes a vital role and with it, the MEDEVAC request.

A proper MEDEVAC request is a hand off of information about the patient and the considerations for care being made both by the team providing the initial care under fire and the responding unit conducting the MEDEVAC. For example, one of my first combat engagements in Iraq was initiated by a mass casualty event. The lead vehicle in a patrol hit a catastrophic IED, amputating both legs of the driver and the left leg of the front passenger, which we call the Truck Commander (TC). Both rear passengers were thrown through the doors and the gunner was ejected from the turret. All were severely wounded. My Platoon Leader coordinated the search effort for the missing gunner while our Medic, with that Platoon's medic, treated the casualties in order of severity. My Platoon Sergeant was busy coordinating the MEDEVAC request over the radio. The MEDEVAC card we all carried with us was the one below:

9-LINE MEDEVAC REQUEST

Line 1: Location of Pick-up Site:

Line 2: Call Sign: **Freq:**

Line 3: Number of Patients By Precedence
_______ A - Urgent (within 2 hrs)
_______ B - Priority (within 4 hrs)
_______ C - Routine (within 24 hrs)

Line 4: Special Equipment Required
A - None
B - Hoist
C - Extraction Equip
D - Ventilation

Line 5: Number of Patients & Type
Litter: Ambulatory:

Line 6: Security at Pick-up Site
N - No enemy troops in area
P - Possible enemy trrops (approach w/ caution)
E - Enemy troops in area (approach w/ caution)
X - Enemy troops in area (armed escort req.)

Line 7: Method of Marking PZ
A - Panels
B - Pyrotechnic Signal
C - Smoke Signal (Color: ______________)
D - None
E - Other

Line 8: Patient Nationality & Status
A - US Military
B - US Civilian
C - Non -US Military
D - Non-US Civilian
E - EPW

Line 8: Patient Nationality & Status
N - Nuclear B - Biological C - Chemical

We located the gunner and successfully marked off the pickup site, getting all of our wounded off the battlefield and into the field hospital. While it was tragic, its a reality of combat. Through using the report format above, my Platoon Sergeant relayed the critical information on the nature of the injuries, what we had done for immediate care, and what special equipment would be required by the responding MEDEVAC team. Had we not done that, the two amputees would have certainly died along with one of the passengers who had suffered a spinal injury. The successful handing off of the information kept them alive.

Not all MEDEVAC requests concern your own troops – once, on my second deployment to Iraq, we were accompanied by the Battalion

Surgeon to conduct a free medical clinic along with the Civil Affairs unit. Two Iraqi men brought an old woman in obviously bad physical condition to us and we immediately knew she was near death from the smell of gangrene. She had had a botched knee surgery that was horribly infected and required amputation. We called in a surgical team, since it was in question whether she was even stable for the flight. The amputation was performed there along with stabilizing care and once her vitals were stable she was transported by air to the field hospital. It is another gesture of goodwill that paid off for our operations in the area - the two men, grateful for her to be alive, provided us with valuable information against the Sunni insurgents in the area.

Training the guerrilla force in MEDEVAC requests may seem mundane but it must be stressed that it is a primary purpose of communications in all three categories, most importantly Sustainment and Tactical. Even if injuries are not combat related, the ability to relay the nature of them and handing off the treatment plan should be a requirement of that training program.

MEDEVAC Request

LINE 1: LOCATION OF PICKUP SITE

LINE 2: CALLSIGN / FREQUENCY

LINE 3: NUMBER OF PATIENTS BY PRECEDENCE

URGENT (IMMEDIATELY PUSHED TO SURGERY)

PRIORITY (WILL SURVIVE 12-24 HRS)

ROUTINE (INJURIES NOT LIFE-THREATENING)

EXPECTED (DOA)

LINE 4: SPECIAL EQUIPMENT REQUIRED

LINE 5: NUMBER OF PATIENTS

LITTER - BORNE:

AMBULATORY:

LINE 6: SECURITY OF PICKUP SITE

N: NO ENEMY TROOPS

P: POSSIBLE ENEMY IN AREA

E: ENEMY PRESENCE

X: ACTIVE ENGAGEMENT / ARMED ESCORT

LINE 7: METHOD OF MARKING CASUALTY COLLECTION POINT / PICKUP SITE

A: PANELS

B: PYRO

C: SMOKE / COLOR:

D: NONE / OTHER METHOD

LINE 8: PATIENT STATUS

A: YOUR COMBATANTS

B: ALLIGNED CIVILIANS

C: FRIENDLY / ALLIED ARMED GROUP MEMBERS

D: UNALIGNED CIVILIANS

E: ENEMY WOUNDED

Clandestine Reports

The other report formats pertain to the clandestine role of communications and the method of sending them covered in Chapter 6. There are certain types of information that must be relayed and these are sent during the communications windows discussed in Chapter 3. In Afghanistan we sent these reports to indicate our team status during long duration missions - typically conducting patterns of life surveillance or interdicting multiple targets over a broad area. In each case we had to send and receive our information from our higher echelon that gave us broad direction during our operation.

These reports originated with clandestine teams operating behind the Iron Curtain in Europe during the Cold War and in Southeast Asia and continues to be used today. It is important to note the names of the reports themselves are meaningless, but included to identify the type of report being sent to the Tactical Operations Center (TOC). The reader will notice they are not labeled Line 1, 2, etc, but rather AAA, BBB, and so on. This has two purposes:

1. So they are not confused with any other type of report
2. When encrypted with a trigram (Chapter 7) are impossible to mistake.

ANGUS Report

The ANGUS Report is sent to report initial entry of a team into an area. The two most dangerous times on a covert insertion by a team is the entry and exfil from the area of operations – a team finds itself the most exposed with the fewest defendable positions from which to fight. For these reasons, during the infil and exfil all radios on a team are switched on. Once they have reached their planned area of operations, they will send an ANGUS report and power down all of their communications equipment to prevent accidental transmissions.

ANGUS: Initial Entry

AAA. Date / Time Group	Date and Time of Transmission
BBB. Team Status	Code Word Identifying Team Status
CCC. Location	Location / encrypted with SARNEG

DDD. Deviations	Any changes reported
EEE. Additional Information	Additional remarks of interest

Example ANGUS Report:

AAA. 22 0345L OCT 2022 (Date/Time and Time Zone/Month/Year)

BBB. GREEN (Team is 100%)

CCC. MD IAAN EHIR (Location is encrypted with current SARNEG)

DDD. Took alternate infil route / planned route had presence of dogs barking

EEE. Village to the immediate west of target has sentries roaming / BORIS to follow

BORIS FOLLOWS // ACK (acknowledge this transmission)

In this report the team has indicated they made it to the planned infiltration site, however the originally planned route was not usable due to the presence of dogs barking. A village that was not originally planned as part of the observation was spotted with sentries and is now a target of interest, and the team is preparing an BORIS report to give a better assessment to the Guerrilla Commander regarding the situation on the ground.

BORIS Report

The BORIS report is a more in-depth intelligence report that relays long term observation and analysis from the team on the ground. It is more broad in nature than what a SALUTE or SALT report contains. The two aforementioned reports are used for immediate or nearly immediate action. The BORIS report is used for long term assessments to coordinate actions among groups over a larger area than what a SALUTE report typically covers.

BORIS: Intelligence Report

AAA. Date / Time Group

BBB. Date / Time of Observed Activity

CCC. Location of Observed Activity

DDD. Observed Activity

EEE. Description of Personnel, Equipment, Vehicles, Weapons

FFF. Team Assessment

Example BORIS Report:

AAA. 22 0345L OCT 2022

BBB. 22 0315L OCT 2022

CCC. MD IANG EHFA

DDD. 7 Armed Sentries / 6 in static positions 1 roaming

EEE. Brown camo pants black shirts, brown plate carriers,

Unidentified patches.

PVS-14 NODs

1 Toyota Truck

Painted M4 w/ ACOG and IR Lasers

FFF. Team appears professionally trained and disciplined. Unaware to our presence. Likely expecting contact from unknown enemy. Will continue observation.

Leader's assessment: attempt should be made in the future to make friendly contact with this group.

NOTHING FOLLOWS // ACK

In this example BORIS report, following the previous ANGUS report, the team describes in detail their observation and includes a professional assessment. Since the element they have observed seems well trained and equipped, but not as of yet hostile, the Leader's assessment is to attempt to make friendly contact in the future.

CYRIL Report

The CYRIL Report is a team status report and is sent at the beginning of each commo window both from the Team in the field and from the TOC as an update for their own situational awareness.

CYRIL: Situation Report

AAA. Date / Time Group

BBB. Current Location

CCC. Medical Status

DDD. Equipment Status

EEE. Supply Status [Batteries, Ammo, Water, Food]

FFF. Team Activity since last Commo Window

GGG. Team Activity until next Commo Window

HHH. Remarks

Example CYRIL Report:

AAA. 22 1545L OCT 2022

BBB. MD IAAN EHIR

CCC. GREEN

DDD. GREEN

EEE. SEVEN BATTERIES AT FULL CHARGE

NO ROUNDS FIRED // 100%

5 GALLONS WATER

12 MRES

FFF. BUDDY TEAM OBSERVED TARGET, NOTHING OF INTEREST NOTED

OBSERVED 7 MAN TEAM PULLING SECURITY, GROUP INITIATED STAND DOWN AND REDUCED SECURITY POSTURE TO THREE SENTRIES. WEATHER PREVENTED LAUNCHING OF SURVEILLANCE DRONE.

GGG. TEAM WILL INITIATE SLEEP PLAN IN HIDE SITE, ROTATE BUDDY TEAMS IN OBSERVATION ROLES. IF WIND SUBSIDES TEAM ANTICIPATES LAUNCHING SURVEILLANCE DRONE

HHH. TARGET VILLAGE APPEARS EMPTY / WILL CONTINUE OBSERVING ADJACENT VILLAGE BUT REQUESTING EARLY MISSION COMPLETION.

NOTHING FOLLOWS // ACK

As noted, the CYRIL report is a detailed status of the team's activities to give the Guerrilla Commander a current status of his team in the field coupled with what their plans until the next commo window will entail. The team indicated exactly how many batteries, how much ammunition, and how much food and water remained. These would normally be sent as a code word to further obscure their meaning.

CRACK Report

CRACK is used to assess battlefield damage to infrastructure, but also has the purpose of identifying whether or not a facility of interest can support the needs of a team on the ground. Traditionally CRACK has been used by forward surveillance teams to assess the conditions of bridges and roads for follow on forces during an offensive. We used CRACK to identify potential safe houses and assess whether or not they'd be suitable for our purposes as well as identify what other resources we'd need to request for extended operations out of them.

In Afghanistan we used CRACK to first identify the location of a potential Afghan Border Police safehouse we'd be using as a staging point for operations along the Pakistan border, then request the assets to improve its capabilities to best suit our needs as well as other considerations. On our first survey we noted the area's natural defenses and how to improve them, then the sanitation considerations, followed by the topography of the area and how it would be suited to signals intelligence collection against the Taliban's VHF signals coming just over the border. All of this was built on a rather large mud hut serving as a transient corral for sheep herders on their way to the Kandahar stock yards. For the potential guerrilla band it can work the exact same in identifying places that can suit the group's needs.

CRACK: Battle Damage Assessment

AAA. Date / Time Group

BBB. Type of Target

CCC. Description of Target (Physical and Functional Damage)

DDD. BDA Analysis / Resource Requirements

Example CRACK Report:

AAA. 22 0345L 23 OCT 2022

BBB. METAL ENCLOSED STRUCTURE

CCC. LARGE ROLL UP GARAGE DOOR

HOLE IN NORTH WEST SIDE OF ROOF

NO STRUCTURAL DAMAGE NOTED

DDD. BUILDING SUITS NEEDS OF TEAM'S REQUIREMENT FOR A TEMPORARY SAFE HOUSE

The CRACK Report is noting a possible temporary safe house inside an unoccupied metal building, also pointing out the damage to the structure while noting its damage poses no threat to the team's requirement.

UNDER Report

The last clandestine report of note is UNDER, or a cache report. While living out of a rucksack is part and parcel of the lifestyle of a guerrilla fighter, we can't carry much, relatively speaking, on our backs for an extended period of time. The lighter, the better, at least when traveling even short distances but over rough terrain. That said life for a guerrilla is very different than that of the counter-guerrilla or an occupation force; one can afford to be overt, the other cannot. We carried all of our gear we expected to use, including water, into the area of operations for extended periods of time. This becomes a finite supply line and a fairly high signature one at that. The guerrilla, reliant as much on blending into a populace as he is with his terrain, can't always have the luxury of a ruck in the open, instead reliant almost solely on caches emplaced by the local support network.

Historical accounts are filled with guerrilla movements relying primarily on these, from the Spanish Civil War to Nazi-occupied France to clandestine activities run by every spy agency on Earth. And in each case, there was a distinct method to marking, and reporting, the location of those caches. The way we would do this is the UNDER report.

UNDER: Cache Report

AAA. Date / Time Group

BBB. Type

CCC. Contents

DDD. Location

EEE. Depth

FFF. Additional Info / Reference Points

Example UNDER Report:

AAA. 22 0345L OCT 2022

BBB. PVC CLEANOUT TUBE

CCC. 210RD AMMUNITION

7 MAGS

AA BATTERIES

3 RADIO BATTERIES

DDD. MD EIRA NARE

EEE. SURFACE LAID / OLD TRACTOR GAS TANK

FFF. MARKED WITH WHITE REFLECTOR TAPE THREE INCHES LONG ON SIDE OF GAS TANK FACING ROAD.

This UNDER report is marking a combat resupply for a team, sealing up a combat load of ammunition, AA batteries for Night Vision and spare radio batteries for an extended operation. It is not buried, but hidden inside the gas tank of a broken down tractor on the side of the road and marked with a strip of reflector tape.

Guidelines for the Formats

The Report formats contained here are a guideline; nothing more, nothing less. If for no other reason they give a glimpse into the way its done in the world of clandestine activities and is by no means an absolute end-all, be-all. It is, however, a logical process by which Western Conventional and Special Operations Forces have worked over a long period of time, condensing that institutional wisdom into a refined product. How your group utilizes the wisdom contained is up to you, but keep in mind, it must satisfy the end goal of handing off the clandestine information in a concise manner. Your life, and the success of your movement, depends upon it.

(This Page left Intentionally Blank)

5. A Crash Course In Field Antennas

The Antenna: The Most Important Part!

When most are first getting into radio they usually ask what make / model to buy, normally asking out of ignorance. The radio itself is mostly irrelevant, aside from its individual capabilities; ie: the frequency range it covers, build quality, etc. But that aside, the most important part of the radio is the antenna itself. Ironically this is also the least understood, with antenna theory falling into one of two categories for those seeking to learn - either too complicated and thus dismissed, or broken down into a series of engineering formulas and added complication looking for an ideal. This chapter takes neither position. The reality is that for a radio operator, even an individual rifleman as part of a larger Guerrilla force, he must understand the performance of his antenna in the same way he does his rifle or any other piece of equipment. His life depends on it.

This chapter explores the antenna first from the individual perspective, the antenna attached to the radio, and then from a higher level of capability with both omni-directional and directionally transmitting antennas for small unit requirements. It is not written from an engineering perspective and the technical data is kept to a minimum so that anyone can grasp the basics of the concepts in an improvised environment.

Individual Fighting Load Radio Considerations

The Baofeng radio lends itself well to a wide variety of roles that satisfies the requirements for Sustainment, Tactical and Clandestine use when certain considerations are made. In its stock configuration, the radio has limited utility and much of that has to do with the generally poor antenna that comes from the factory. Most buy aftermarket antennas along with the radio.

The Baofeng uses a SMA female adapter, which is small, generally fragile, and like all threaded connections, has a limited life span. It also is structurally weak when used with larger aftermarket antennas which leads to them breaking under hard use. This absolutely must be a consideration for taking these radios into the field and the same is true for any other radio utilizing SMA connections. I convert all of these to BNC, which is far more robust for field use. BNC connectors have the added benefit of being twist on / twist off, making quick connections secure and simple in the field. This makes attaching the antennas we will cover in this chapter far easier. My personal choice in an inter-team, tactical level antenna is anything that is highly flexible and gives good performance both in transmission and reception.

Running a radio on your personal fighting load necessitates placing it on your non-primary side to avoid striking the antenna with your rifle stock. I weave the antenna in the webbing of my shoulder straps. I do not advocate using an antenna relocation kit as is popular with some - these are fragile and create an unpredictable radiating pattern due to the interference of the body's electrical field and the interruption between the antenna elements discussed later in this chapter. For troubleshooting reasons a rifleman is much better off with a conventional antenna setup.

Aftermarket antennas for a handheld radio are only going to get a team so far, however, for a large number of reasons. To be used in any role other than the tactical, and specifically the inter-team level, building your own field expedient antennas becomes a requirement. The remainder of this chapter is dedicated to instructing the underlying theory on exactly how to do that in simple terms, granting the reader the tools to build any of the common antennas required for maximizing the capability of the Baofeng (and any other radio) in an austere environment.

Understanding Radio Theory

One of the most common questions regarding radio in general is 'how much range' does it have? This is a problematic question, because, when pertaining to the Baofeng and any other handheld, or even mobile VHF / UHF radio, to give a proper answer it requires an explanation of basic radio theory. The answer in layman's terms is, universally, that it depends on a variety of factors:

1. The Operating Environment
2. Obstacles between you and your intended receiver
3. The efficiency of your antenna

It is critical to understand that first your handheld radio operates in the VHF and UHF ranges, which is best understood as Line Of Sight (LOS). What this means is that, in theory at least, if one radio can ‘see’ another on the radio horizon, it can communicate with it, regardless of any other factor.

One of the general rules concerning handheld radios is that the antennas themselves are usually a compromise - just efficient enough to keep the radio from burning itself up during transmission and giving just enough signal quality to get a couple of miles, if that. Commonly users change the stock antennas out for aftermarket models that give a little bit of improvement in some cases. We will address this at the inter-team tactical level. But with that said, the best antennas are purpose built for the task and this chapter lays out basic instructions on how to do so with improvised equipment in any environment. This will enable the Baofeng (or any other radio equipment) to have far more capability than most believe possible, while aiding in the Communications Security (COMSEC) requirements of Tactical and Clandestine roles.

In order to understand improvised antenna construction we need to take into account the basics of physics in layman’s terms. **Radio waves travel at the speed of light.** With that said, it is best to visualize radio waves, and radio propagation, as a form of light. Do this - imagine yourself in a dark room. There’s a lightbulb that suddenly comes on. That lightbulb is your radio and antenna when its transmitting, and everything that is illuminated is where your radio waves are propagating, or can be heard. If you can visualize this picture, then you already have a basic understanding of radio theory.

Environmental Impacts on Radio Propagation

The operating environment makes a huge difference in the physical distance that a radio wave will travel. Regardless of one’s purpose for establishing communications, be it Sustainment, Tactical or Clandestine, the operating environment mush be taken into account to recognize the potential limitations and assets present. As we covered briefly in Chapter 1, the general rule is that VHF signals (30-300mHz) perform better in rural, hilly environment and that UHF (300-3000mHz) performs better in urban sprawl. This is due to VHF ground wave signals physically bending with the terrain - keeping with our lightbulb example above, small terrain features such as hills cast a smaller shadow in that illuminated room. They also refract a signal off those same terrain features in the same way light does in a mirror, scattering your signal. Conversely, UHF signals are blocked by even minor terrain features but penetrate buildings. These basic facts must be taken into account when writing a tactical SOI.

With regard to COMSEC, a team may want to use UHF for inter-team communications and VHF to send messages back to the TOC. The inter-team signal does not need to travel far, maybe only a few hundred meters at most. This mitigates potential interception. Communications with the TOC, or communications with the Guerrilla Commander, usually covers a significantly longer distance, and VHF in conjunction with a purpose built antenna makes sense when operating in a rural environment. If operating in an urban environment, the reverse would be true.

There are several other environmental impacts on radio waves aside from the general rules laid out above and in Chapter 1.

1. **The higher the frequency, the more vegetation degrades signal.** In a jungle environment expect a radio signal coverage range in terms of physical distance, depending on frequency, to be

shortened by at least half or possibly more. When conducting the RTO Course in the spring in woodland environments VHF universally performs better. The US Army study of the PRC-64 during Vietnam contained a table calculating loss at various frequencies in a Jungle environment. HF performed the best, but is mostly impractical due to large antenna size. VHF was the best compromise for tactical operations. UHF performed the worst in terms of physical coverage distance, but presents an advantage for communications security.

2. **Desert environments reflect signals over a much longer range than usually expected.** Due to a lack of vegetation and a higher degree of soil reflectivity, radio signals cover a longer physical distance than in woodland or jungle environments. Performance in Arctic environments, mostly snow and ice covered, perform similarly.
3. **Bodies of water provide a perfect reflective surface.** For the maximum amount of coverage, the ability to transmit over a body of water provides the perfect reflector for a radiated signal.
4. **The better reflective properties of the soil, the higher the quality of your transmission.** In simple terms, soil that has higher mineral deposits of silicates, granite, etc, perform much better in terms of radio wave propagation. If there is poor soil at the transmission site, adding water underneath the antenna will improve it. Frequently the communications specialists would urinate under the antennas to improve their reflectivity.

Obstacles between You and the Recipient

The second factor affecting the range of a radio are physical obstacles between you and the intended receiver. The Baofeng, and any other radio operating in the VHF / UHF bands, are known as Line Of Sight (LOS) means of communication. If one radio can 'see' the other in terms of radiated energy (known as RF), they can communicate. Obstacles between you and the intended receiver can prevent this. Large hills, ridge lines, and urban sprawl all present the most challenging obstacles in a given environment.

The way this is resolved is by increasing one's physical line of sight. This is the reason that in mountainous terrain you'll see antennas along the highest ridgelines and why antennas are usually found on top of tall buildings in urban areas. **The more line of sight, meaning the higher in physical elevation you go, the more physical distance a radio wave can actually cover**. This is also known as groundwave. The higher both you and the intended receiver are over the physical horizon line, the longer both of your lines of sight will be, respectively and thus the more physical distance you will cover with your radios regardless of power output.

Obstacles in our environment also scatter our signals through refraction. Hills and building both reflect a signal much the same way light reflects in a mirror. This plays a big role in clandestine communications techniques with directional antennas covered in Chapter 6.

Antenna Efficiency

This brings us to the big question regarding range – the antenna itself. When most think of antenna, they think of the thing sticking out of the top of the radio. And while that is part of the antenna, its exactly half of the antenna. The external part of the antenna may get all of the attention, but it ignores the metal body built into the radio also known as the ground. **All antennas are a form of a dipole**. If you can understand the concept of a dipole, then understanding antennas can become simple.

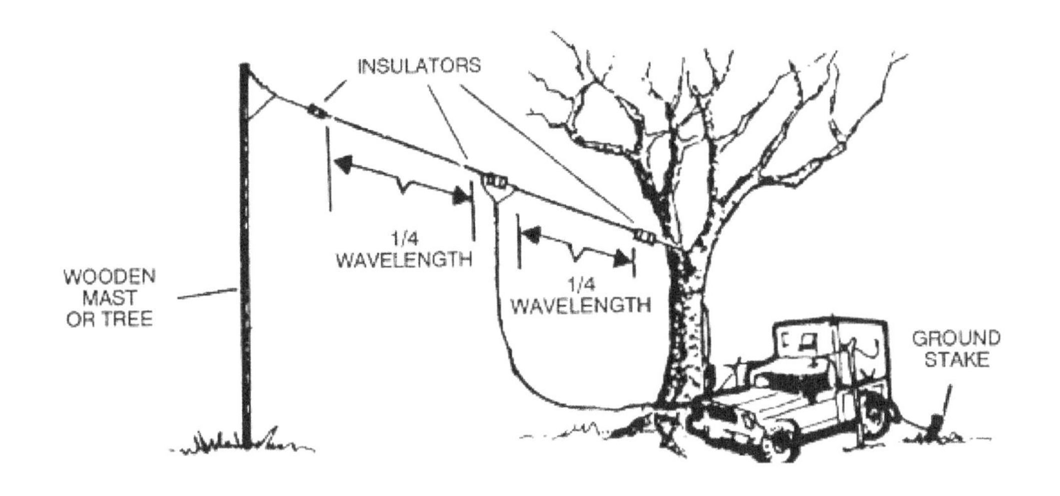

A dipole antenna has two equal and opposite halves – a positive and a negative. In a handheld radio the physical antenna is the positive half and the built in ground is the negative half. This is also referred to in engineering manuals as the reflector. Both sides of the dipole are considered mirror opposites of one another, and the closer they are to matching one another the more efficient the antenna. The optimum antenna length, once calculated, is also known as **resonant**. A resonant dipole gives 2.15db of gain – in layman's terms, db of gain represents how much stronger a signal is in a similar manner as candlepower or lumens of light is understood.

All antennas have either gain or loss, and the further an antenna is from being a true or resonant dipole, the more loss. 3db of gain doubles your effective radiated power output; put simply, it makes a signal be as strong as if it were transmitted with double the power. So understanding that a dipole represents 2.15db of gain, in building a better antenna we can nearly double our signal strength without sacrificing more power. Gain works in orders of magnitude. With every additional 3db of gain, we double the power again.

The physical length of the antenna must be calculated and this is done with a simple formula:

FULL WAVELENGTH: 936 / Frequency in MHz = Physical Length in Feet

936 represents the mathematical constant for a full wavelength of RF energy. Antennas, however, do not operate in full wavelengths. Dipoles are a half wavelength long. Why is this?

Back when I was learning this in the Army I asked this question to one of the commo instructors and was quickly told to simply stick to the course material. I never learned the answer until much later when I had a physicist in class. She explained that RF energy traveling up and down antennas, like jump ropes, work in a parabolic path. If the rope is too short, it will move too fast, leading to inefficiency. If its too long, it will be just as inefficient and possibly even worse than being too short - that a half wavelength represents the ideal length in efficiency. A half wavelength was the ‘Goldilocks’ - just right.

The formula for calculating a half wavelength dipole is simple; divide 936 by 2, giving us 468. The formula for calculating the total length of a dipole becomes:

HALF WAVELENGTH: 468 / Frequency in MHz = Antenna Length in Feet

VHF example: 468 / 151.820 = 3.08 ft

UHF example: 468 / 462.550 = 1.01 ft

This leads us to the individual pieces of the antenna, also known as elements, that we need to measure and cut for the two halves of the dipole antenna. To do that, we divide 468 by two once more to give us 234. Our formula for each antenna element becomes:

QUARTER WAVELENGTH: 234 / Frequency in MHz = Length in Feet

VHF example: 234 / 151.820 = 1.54 ft

UHF example: 234 / 462.550 = .506 ft (6 inches)

As a rule, I always check my work by first calculating the quarter wave lengths and adding them together, then calculating the half wave length to see if they come out the same. This way, if there are any errors in my math, I can figure out where the issue lay before I waste any material.

Notice the antennas get substantially smaller the higher in frequency you go. Antenna physical length is inversely proportionate to frequency, meaning the lower in frequency, the longer the antenna, and vice versa. While UHF may not be the best performer in rural environments, its antennas are small, meaning a lower signature and more portable. Certainly something to consider.

Antenna Wire

Any type of wire can be used for antennas. This makes sourcing wire in an austere environment fairly easy, both in rural and urban areas alike. There are however, a couple of considerations to make.

1. Bare ware can ground itself out when touching vegetation
2. Solid wire has a tendency to break or back out of crimps

I'm a frequent user of electric fence wire for antennas; its cheap, plentiful, and disposable. In saying that, it has an inherent weakness that is pointed out above - being a bare wire, if it touches anything grounded it becomes grounded itself and can damage the radio. It also has a tendency to break or back out of its own crimps due to being solid wire.

Other types of wire to consider are THHN, or stranded electrical wire used in construction, speaker wire and lamp cord. While significantly more expensive, they are made of tiny strands bundled together and coated with either rubber or plastic as an insulation, making them more durable in the field.

We routinely made field expedient antennas out of Claymore wire, which is nothing more than aluminum stranded lamp cord. We did the same using WD-1 commo wire, also known as slash wire, which is used to connect field phones. It is also made similar to lamp cord but far lighter. In my experience the lighter the wire, the more prone to breaking. 16-14 gauge wire is ideal for tensile strength versus weight.

Insulators

On each end of the antenna elements (the pieces of wire), we attach insulators. An insulator terminates the end of the antenna element, providing a place to attach securing lines or ropes for hoisting. **550 is not an insulator.** When wet, it becomes electrically conductive. Do not use it as an end insulator.

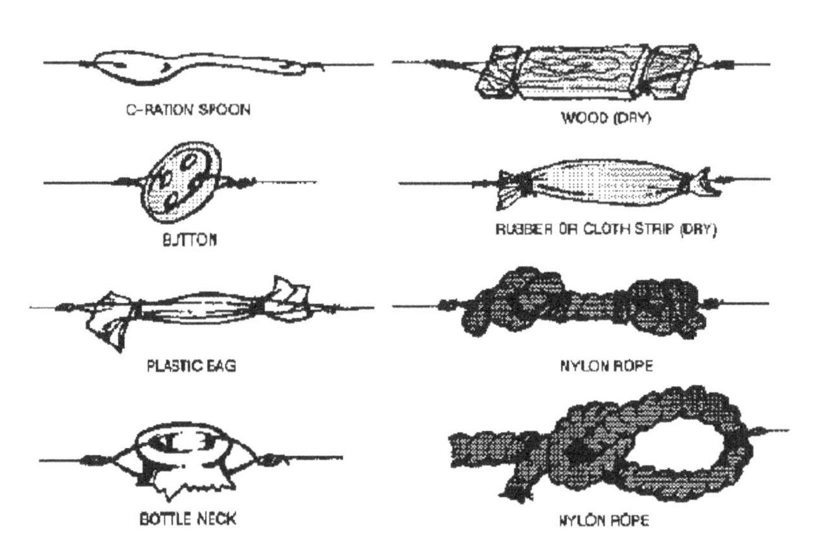

Field expedient insulators are easy to find in literally every environment I've operated in – whether its pieces of plastic or glass from bottles, pieces of wood in desert environments (this is not ideal in humid areas or jungles), plastic spoons with holes cut in them for the wire and the securing line, or my personal favorite, electric fence insulators made of either plastic or ceramic. They do not have to be expensive or pretty to be effective – they're field expedient.

We attach the insulator to the wire by bending a loop (known as a bite, just like in mountaineering) and attaching the insulator at the end, completing the loop and either bending or crimping it in place. The RF energy will travel to the apex of the loop rather than the end of the wire.

Antenna Orientation / Polarization

Antennas of any type are polarized based on the orientation. Up and down is known as **vertical polarization**, parallel with the surface of earth is known as **horizontal polarization**. Handheld radios, and most line of sight radios for that matter, are vertically polarized - meaning the antennas are straight up and down. There is a 12db difference in signal strength between changes in polarization - meaning that when you go from vertical to horizontal, a vertically polarized antenna may or not pick up your signal. This is important to note for COMSEC considerations. If all assets dedicated to finding your signal are vertically polarized and you are horizontal, they may miss you entirely.

On the other hand, there is a reason for vertical polarization with most LOS radios. Since they operate in Frequency Modulation (FM), the proximity to the earth ground itself can lead to the scattering of signal and induce loss. This is not always the case, as noted in the environmental concerns above, but as a general rule to prevent those issues FM signals generally use a vertical polarization. Again, recognizing this fact can be used to your benefit.

Field Expedient Antenna Designs

Whether the role is for sustainment communications or for extending the communications range of a patrol sending and receiving transmissions from a TOC, the field antenna designs covered here are simple in construction and high in performance. These antenna designs fall into one of two categories: **omni-directional** and **directional**.

Omni-directional antennas radiate RF energy in all directions simultaneously. Image yourself in that dark room once more - a lightbulb comes on and lights up the room equally in all directions. **Directional antennas, on the other hand, perform like flashlights, radiating RF energy in the primary direction in which they are pointed**. Just as how a flashlight illuminates only where its beam is pointed, the same concept applies to the directional antenna. Directional antennas have a strong benefit to COMSEC in that unless a receiver is along the radiated path, interception will be nearly impossible. Further, directional antennas have high gain values, making them very efficient for extending the operating range along the directed path. Just like how a flashlight illuminates an area its beam is pointed stronger and further than an omni-directional light source will, the directional antenna does the same.

The Jungle Antenna

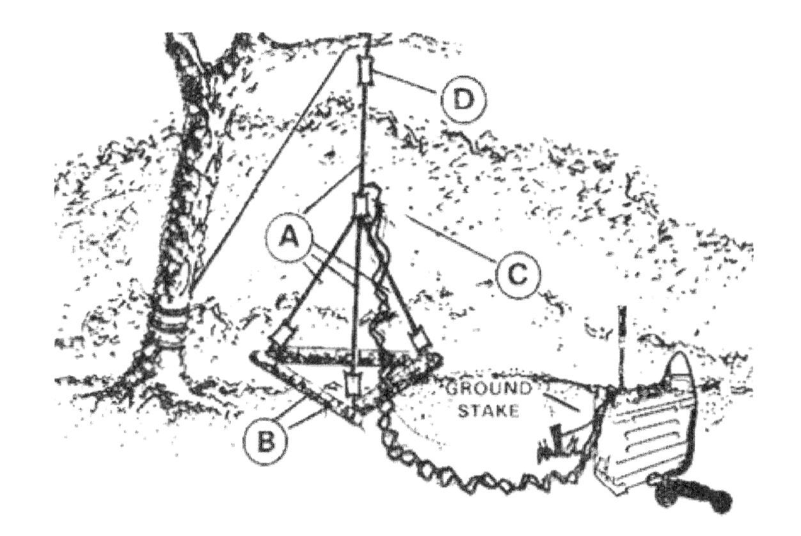

The Jungle Antenna, also known as the 292 Antenna, the groundplane antenna, and in the military the OE-254, is a simple to construct, vertically polarized, field expedient antenna that gives us 6db of gain in all directions. It gets its name from the pre-WWII era when US forces were training in jungle combat for a potential war in the Pacific. Radio was a relatively new implementation on the battlefield since WWI and with the advances in technology came the requirements to best implement it. While training on various means in Panama the US Army Signal Corps discovered that adding two additional legs to the negative side of a vertically polarized dipole, forming a type of pyramid, performed well in heavy vegetation.

One of the easiest ways to construct the Jungle Antenna, and any other antenna, is to have an attachment known as a **Split Post BNC Adapter**, also known as a **Cobra Head**. These have a positive (hot) and negative (cold) terminal marked by red and black screw down caps, respectively. These make attaching coax cable and running it to the radio simple. To construct the Jungle Antenna, cut four Quarter-wave lengths of wire based on the operating frequency. One will attach to the Hot end, the others will attach to the Cold side, forming a pyramid. Frequently people cut three sticks and tie them together forming a triangle base for the cold end. This becomes the groundplane.

Next, attach the coax cable. Most transceivers, aka two way radios, use 50 ohm coax. For field expediency, any 50 ohm coax cable will work fine. Understand that there will be some loss, but for these purposes that will not become a major factor unless extremely long runs of cable are used. Under 18 ft, you'll be fine. The two most common types of cable I carry are RG-58 and RG-8X, both fitted with BNC connectors making attachment to the radio fast, simple and secure.

The Jungle Antenna's use is for creating the maximum amount of signal range from a fixed location. For sustainment communications purposes, this would be extremely useful for creating communications during the recovery phase of a natural disaster. This extends the RF range to roughly 6 miles. For tactical purposes, this radio would be in a fixed location at the TOC and for a receiving at a dedicated transmission site on a patrol as discussed in Chapter 6.

Directional Antennas

Directional antennas, as described previously in the chapter, transmit along the direction its aimed known as an azimuth. On the map of the operating area, the Guerrilla force should first mark the location of the TOC or Safe House, then the planned transmission sites in the area of observation. Once they've done so, they get the azimuth from the TOC to those planned locations and the back azimuth (subtract or add 180, 360 degrees in a circle) from the planned transmission site to the TOC. Using the directional antennas, they'll send and receive radio traffic in the report formats covered in Chapter 4 through the encoding / encryption covered in Chapter 7. This mitigates the chances of interception and direction finding (DF) of their transmission, compromising the team.

The Sloping Vee

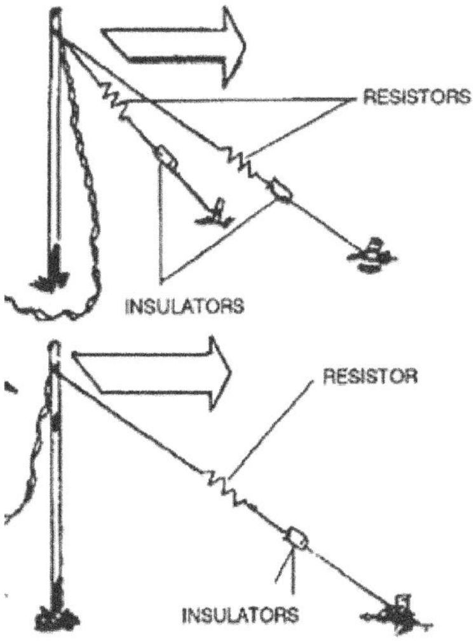

The Sloping Vee is the simplest directional antenna. It is a Dipole with the two legs brought into a 45deg angle from one another, with the opening of the angle pointed in the direction of the intended transmission. It is quick to construct and hoist, also making rapid adjustments easy.

RESISTORS
INSULATORS
RESISTOR
INSULATORS

The diagram includes resistors, however this is not required. The addition of a carbon resistor of 500 ohms or more maximizes the amount of RF energy being pulled forward, but the sloping of the legs of the dipole already creates a radiating pattern moving forward in the intended direction.

The Yagi

Probably the most recognizable directional antenna and one of the most practical ones is known as the Yagi, named after one of the two physicists that designed it for Imperial Japan prior to WWII. The

Imperial Japanese Navy was exploring primitive forms of RADAR that the British would later perfect by directing RF energy along a path and listening for a ping, or whatever reflected that signal back to the source. The Japanese also recognized this antenna’s utility for communications security, noting that when aimed along a specific azimuth interception of the radio signal outside that path was made extremely difficult. This would become their technique for communicating between the command posts of each occupied island in the months leading up to the Pacific theater of WWII.

After the war the Yagi would become nearly ubiquitous in the US with the proliferation of over the air TV. Most households would have a Yagi antenna fixed to the roof with a dial, controlling a rotor, turning the antenna in the direction of the TV station for the best reception. If you can understand that concept, you already understand how the Yagi works.

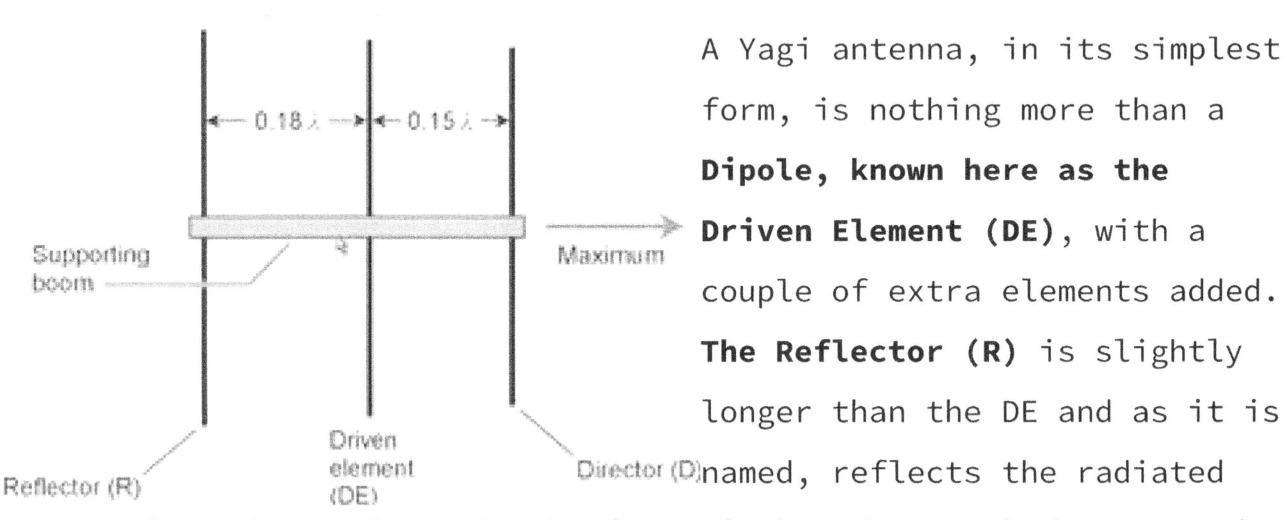

A Yagi antenna, in its simplest form, is nothing more than a **Dipole, known here as the Driven Element (DE)**, with a couple of extra elements added. **The Reflector (R)** is slightly longer than the DE and as it is named, reflects the radiated energy from the DE forward. The front facing element is known as the Director (D) and pulls the radiated energy forward as well. The Director functions the same way as the front post on an aiming sight. This Yagi is known as a three element Yagi and has a gain of 7.5db.

Spacing between the elements are magnetically related. They cannot be too close or they run the risk of becoming parasitic on one another, but also cannot be too far away, or they lose the properties required to make them work with one another. For that reason, the spacing (labeled in fractions of a wavelength) are given on the boom, or the insulated piece running down the center of the antenna structure, in the diagram. To calculate this we need to first calculate the mathematical constant for each length. We do this by multiplying those fractions of a wavelength by 936.

BETWEEN R AND DE: .18 WAVELENGTH (936x.18=168.48)

BETWEEN DE AND D: .15 WAVELENGTH (936x.15=140.4)

The constant for the boom length between R and DE is 168.48, the constant for the length between DE and D is 140.4. The constant for the total boom length becomes 308.88. To get the physical length, we go back to our formula for calculating the length in feet.

Total Boom Length: 308.88 / frequency = Length in Feet

R to DE: 168.48 / frequency = Length in Feet

DE to D: 140.4 / frequency = Length in Feet

I always start by calculating the total boom length necessary, then cut a section that length. Any insulating material works well - I prefer small diameter PVC or fiberglass fencing rods, but wood works as well when its dry. Once I calculate the total length needed, I work from the back to the front, marking off the placement of the Reflector first, then the Driven Element, then the Director.

Once that's done, now its time to cut wire. Copper welding rods also perform extremely well for building Yagis.

R: 510 / Frequency = Length in Feet

DE: 468 / Frequency = Length in Feet

D: 425 / Frequency = Length in Feet

Again, the DE is nothing more than a Dipole. For a fast and easy construction, I use the Cobra Head as the feed point (where the coax attaches to the antenna) and use electrical tape to secure the Yagi elements in place. The R and D elements are single pieces of wire. It works.

This Yagi, prepped and ready for use with a camera tripod, was constructed in the RTO Course out in Wyoming. We achieved a 35 mile contact via voice and a digital data burst from this point to another Yagi built in class positioned on a distant ridgeline.

The Yagi antenna can be added to; more directors can be added to the front of the boom, with the calculations for spacing and length being the same. A 5 element Yagi produces 9.5db of gain and has a tighter beam width on the transmitted signal. But that said, it also receives signals that much stronger, and runs the risk of deafening the receiver in the radio. Think of it as someone shouting in your ear as loud as they can versus having a normal conversation – when your radio receives that much stronger, it is the same effect. The 3 element Yagi is highly effective and for that reason I stick with the basics.

A word on SWR

Field expedient antennas are exactly that – built from commonly available parts to get the job done. They're not always pretty, but they work as well as, or in some cases greatly exceeding, a commercially built piece of equipment. That said, attention needs to be paid to the Standing Wave Ratio (SWR), the measure of the antenna's efficiency. The SWR is a ratio of how much energy is being sent forward versus how much is reflected back to the radio. A perfect SWR is 1:1. Anything below 2:1 is usable. Anything above 2:1 runs the risk of damaging your transmitter by built up heat returning to the radio and burning the final transistor. You can measure this with a dedicated SWR meter that I strongly suggest a Guerrilla support group have in their tool kit.

The Baofeng radio does not have built in SWR protection, so when building your field expedient antennas, attention must be paid to the measurements of the elements themselves. The more precise, the closer to perfect resonance, which in turn is closer to that 1:1 SWR. But above all, the closer we are to a 1:1, the more efficient our transmissions are, which could mean the difference between life and death.

6. Communications Operations

The three identified roles of communications - sustainment, tactical and clandestine, each have specific considerations for their respective operations. It is up to the Guerrilla Commander and those tasked with communications to recognize the role at any given time. Failure to recognize the importance of this principle is the first and likely most devastating failure concerning COMSEC.

Individual Considerations

Communications equipment must be limited to key leaders and those tasked with communications as a specialty skill. COMSEC is every bit as critical as understanding the nature of communications itself. A Baofeng can save your life, and when undisciplined in use, can get you killed. Radios of any type must be understood the in the same way as discipline with a rifle and mitigating noise and light while on patrol. In Afghanistan, Iraq, Ukraine and the battlefields of the future great pains have been made to monitoring the electronic spectrum, and the best advantage a Guerrilla has in the face of a technologically superior adversary is to limit the transmissions themselves based on the principles already discussed.

Leaders have a reason to communicate, as do their specially trained communicators tasked with carrying a radio. Erroneous or unnecessary traffic should be kept to a minimum. In all cases, the message to be sent should be written out before transmitting over the radio. This is called writing a script, preventing errors such as the over use of ‘um’ when spoken or errors in transposing the message in a digital format. In following these basic practices, traffic is kept short and to the point.

Sustainment Communications

Sustainment Communications, as identified in Chapter 1, are used to establish conventional communications where there otherwise would be none. This could be after a natural disaster or creating an ad-hoc network when an oppressive government shuts communications down.

Nigeria

When Boko Haram began its terror campaign throughout Northern Nigeria in 2012, it first attacked the cell phone towers before any other target. This served two purposes - discredit the Nigerian government's ability to protect its own critical infrastructure and second, to create area access denial the region's Christian population. Once cell phone service became unreliable, the local populace quit relying upon it, shutting off their main artery for information. Their own social networking slowed down, allowing Boko Haram to exploit this reactionary gap, attacking Churches and kidnapping women from Christian schools to be sold into slavery.

The government's response first had to focus upon consistently restoring and protecting those communications in an attempt to isolate the budding insurgency. By the time they had successfully done this, due to the other infrastructure and social challenges Nigeria faces, the insurgency had moved on to more complex targets. This could have been circumvented with a trained populace in communications skills.

China

During the Hong Kong freedom protests of 2018, and ongoing protests in reaction to COVID 19 social isolation measures, the Chinese government moved in to cut conventional communications from the areas in open revolt. Since the majority of the protests were being coordinated via social media, the internet nodes, run through the cell networks, were shut down.

Even with the expansive ability to monitor social media by any competent government, the reactionary gap still exists that is exploited by the potential insurgent. The protesters in this case were using it in two ways - to rapidly coordinate flash mobs that would gather then disperse before a governmental response can be alerted and second to generate false traffic to divert that same reaction, creating yet another reactionary gap. This was a technique explored by protesters fomenting color revolution during the Arab Spring.

The advantage radio has over its more sophisticated, cell phone based means is its decentralized nature. A network of Baofeng radios, absent any other required infrastructure, can be used to coordinate these activities with little training. Despite the surveillance capacity, it creates an even broader reactionary gap in which a Guerrilla can operate.

Operation Cuba Libre

Che Guevara, Camilo Cienfuegos the Castro brothers and the rest of the 26th of July Movement landed near Los Cayuelos, Cuba in 1956, beginning their ultimately successful guerrilla campaign. One of the first tasks was to set up a radio station, Radio Rebelde. This would broadcast favorable propaganda to the Guerrilla band and coordinate their activities over a broad area in the Sierra Maestra and ultimately the entire island. While they lacked many of the capabilities described in this manual, it is a good example of radio in a sustainment role being successfully implemented.

In July of 2021, mass protests erupted in Havana, Cuba in opposition to the continuing oppression and poor quality of life provided by the communist government of Miguel Diaz-Canel. Internet infrastructure on the island nation is limited and tightly controlled, as is cell phone access, all of which is pipelined through the Directorate of Intelligence (DI). Regime change for freedom in Cuba has long since been an American goal, with the secondary goal of reducing influence from Russia and China in stabilizing their government. Amateur radio, however, is very much alive and well on the island and is a popular hobby in addition to being a critical resource for disaster relief from hurricanes.

When the protests began, many of those same radios began to play a vital role in coordination. Unfortunately much of the equipment was either in the HF or low band VHF portions of the spectrum, as the government highly restricts the sale of handheld VHF / UHF radios, including the Baofeng. The flow of information was slow, creating a very narrow reactionary gap that was dependent more upon interpersonal contact rather than electronic. The Cuban government shut down all amateur radio transmissions and shut down Radio Havana as well. The island would go radio silent.

The Radio Recon Group took to the air to transmit messages of freedom in solidarity with the plight of the Cuban people over the amateur airwaves, in all cases eventually being jammed by strong radar signals originating from Havana on those same frequencies. Had the protesters had access to Baofeng radios and an independent communications infrastructure, they would have had a far better chance at success.

Sustainment Communications for the Guerrilla

A Guerrilla Commander's primary concern is not combat. It is maintaining the favor of the people. The guerrilla band exists as a means social reform in the face of oppression or injustice. As has been described in the historical examples, sustainment communications plays an incredibly important role in a guerrilla war.

Configuring the Baofeng for this role is simple – it is a conduit for information where there may be no option otherwise. Using a standard SOI specifically for this role and with an omni-directional antenna, you can create communications for a local area quickly. One of the first tasks, after getting the antenna on the air, is to check the physical reception range. This is done by creating physical distance between radios and marking their locations on the map. Generally speaking, a Jungle Antenna with a Baofeng will realistically give a range of 6-10 miles, terrain and frequency dependent. Repeaters can also be utilized in this role to extend the operating range as described in Appendix A. These should not, however be relied upon for tactical purposes. Analog repeaters were frequently used in the tactical role by the Taliban, were easily compromised and often destroyed, cutting their communications capability. This led to them making heavy use of HF radio instead. The spare Baofeng radios should be dispersed among house holds along with the SOI specifically for that local net, creating the necessary infrastructure independent of a grid.

Not only does this generate favor among a local populace for the Guerrilla, it also serves as a means to provide medical aid where needed, information for targeting forces more effectively, and improving morale over a targeted area, all of which works to foster a positive influence among the local populace and the Guerrilla.

Tactical Communications

Using the Baofeng in a tactical role has a substantial number of concerns for preserving COMSEC. The two advantages a Guerrilla maintains is his speed and surprise. Undisciplined use of a radio compromises both. As stated, only key Leaders should be assigned radio equipment in a tactical role. This keeps unnecessary traffic to a minimum.

Tactical radio communications serve two distinct purposes: **coordinating fire and maneuver**, and, **relaying battlefield information**. Radios are used by the Leaders on a patrol to talk to one another and coordinate movement. The reality is that small teams, when dispersed over an area, and within dense vegetation or terrain features, usually end up outside of visual range from one another. Recognizing that most movement conducted at night, this requires using radio communications for everything from preventing fratricide during a react to contact to coordinating a support by fire position and an assault element on a raid.

When using a Baofeng in the Tactical role:

1. **Keep your transmissions short. Use codewords whenever possible.**
2. **Always operate with a separate receive and transmit frequency.**

3. **Cut the power to the lowest level possible and use a shortened antenna.** This cuts the physical range of the transmission down, mitigating the interception and direction finding (DF) threat.
4. **Do not transmit on top of terrain features.** If at all possible, remain in valleys or no higher than the military crest of a ridge line. This masks your signal and the terrain also reflects it, making DF difficult.
5. **Use the opposite band than what is optimum for the terrain.** If you're operating in rural environments, use UHF. If urban, use VHF. The idea is that by using the least optimum frequency range, you're limiting the physical distance your radio traffic can be heard.

The role of the TOC is to coordinate and direct the activities of the tactical elements on the ground. The TOC must be flexible and will not necessarily remain in a static location at all times. In Afghanistan, during large scale extended operations, the Ground Force Commander ran a mobile TOC from his vehicle, coordinating our activities spread out over several miles. We had multiple mountains and ridge lines between us, but his group served as his TOC while he reported to a higher echelon. A Guerrilla TOC functions much in the same way. A TOC requires:

- A Radio to communicate with the Teams or Maneuver Elements
- An omni-directional antenna to maximize the radio range
- A way of recording and displaying reports
- A map of the area of operations
- A coffee maker

The TOC will have an omni-directional antenna, maximizing the range of their RF signal as well as having the strongest reception capability possible in any given direction. However, unlike the Afghanistan example, a Guerrilla must recognize the threat of interception and their status as a high priority target. For that reason, careful SOI planning and adherence to COMSEC procedures need to be reinforced.

Clandestine Communications

Communications of a clandestine nature often initiate tactical action. This is either a result of a directive based on intelligence collected or in reaction to a compromise. Clandestine communications are used by teams in the field to relay information collected based on the report formats listed in Chapter 4. When applied to intelligence collection, they are known as **Products**.

These types of transmissions are normally sent over longer distances and during the times described in the communications windows. This requires the use of directional antennas. In the mission planning phase a patrol, no matter the size or scope, will identify their transmission sites and record the azimuth to the TOC. This will be used to transmit the encoded / encrypted messages using the techniques described in Chapter 7.

As a rule, no radio transmissions will originate from the Hide Site, Patrol Base, or Safe House. Transmission sites must be located at least 1000m away from those to preserve the operational security of the team. Those transmission sites are planned in this manner to avoid additional casualties if the transmission itself is intercepted, direction found (DF), and artillery is launched on the source. This became standard practice among Unconventional Warfare (UW) units in the Cold War era.

When selecting a transmission site, its a good tactic to place it low on a hill to the rear of the intended transmission azimuth. This uses the terrain feature itself both as a reflector but also as an RF shade to mitigate DF. When teaching the RTO Course in Wyoming we used the class location, situated near the base of a major mountain ridge line, as a type of backstop for our signal. Using Yagi antennas from our primary location to the other that we had planned, we successfully burst a data transmission and established voice communication at a range of over 35 miles on 4 watts. Going back to our light wave example, we were essentially aiming two flashlights at one another in a dark room.

Remaining low in the terrain also scatters your signal to those with higher line of sight, including aircraft. In Utah we noted that when communicating point-to-point in the low valleys, the team tasked with signals collection had an almost impossible task even when they climbed an overlooking mountain. The directional transmission made reception difficult enough, when coupled with the scattering effect of the terrain itself, it became nearly impossible even with highly capable signals interception equipment, such as a high end Software Defined Receiver (SDR) and a tuned antenna. The team tasked with communications, however, had no issue.

The equipment fielded by a government backed occupation force will have no additional capability beyond that of what is currently common / off the shelf and in many cases exceeds it. So while this not intended to make anyone complacent by any means, it is meant to underscore the reality of physics and the utility of the techniques itself, rather than to give way to abject and largely unbased fear.

The reality is thus that in utilizing all of the techniques described herein, an armed force can be well supplied and highly capable with inexpensive equipment.

7. Digital Encoding and Encryption

The Baofeng radio transmits an analog signal - meaning, in short, that there is no built in encoding or encryption capability to the device itself. But that is not to say that this cannot be done. Further, as radio equipment and communications have become more complex over the past five decades and into the future, many overlook the capabilities and advantages of analog radio along with ignoring completely the old, tried and true methods for encrypting messages for the maximum level of security.

Encoding versus Encryption

Encoding is not synonymous with encryption. One simply takes an otherwise obscured message and transposes it into a code. This offers a level of security to anyone who does not possess either the means to recognize the code nor the capability to decode it. **Encryption** is the intentional obfuscating the contents of a message to hide its contents. One is not the same as the other, but it is important to note that when the two are mated we create proper communications security. Encryption comes in two broad forms: **Digital** and **Physical**.

Digital Encryption relies on software built into the device itself to encode then encrypt, and in turn decode and decrypt, whatever is sent over the air in that particular protocol. Examples include VINSON and ANDVT which is built into US Military radios and AES encryption that DMR utilizes. **Physical Encryption** encrypts the message by hand. This is the most robust method of encryption and can be sent via any means, including analog voice. Physically encrypting messages becomes a critical task to communications coordinating Guerrilla activities.

Move / Counter-Move

At the time of this writing, many of those lessons are being re-learned in the Ukraine-Russian War. Much of the communications equipment is commercial / off the shelf, with both Digital Mobile Radio (DMR) and analog Baofengs being fielded. Early in the conflict DMR showed advantages due to the built in encryption capability and the ability to burst short SMS messages. But for every move there is a counter, and the Russians countered by first limiting their own side to analog only, leaving the only DMR signals on the battlefield to the Ukrainian side. Once a DMR signal, which is highly unique both in its sound over the air and how it looks on a monitoring spectrum, was identified it immediately became a target of interest for artillery and rocket fire. This would in turn lead to both sides returning to analog only transmission.

This cat-and-mouse game is the most recent example of any and every technique having a shelf life - once a pattern is set, a pattern becomes recognized, and the craft of intelligence and its value in predictive analysis is based on pattern recognition.

Encryption is unlawful in Amateur Radio. With the exception of configuring a Baofeng radio for use with a tablet and a freeware app known as andFLmsg, the techniques discussed in this chapter are not meant for Ham radio and should not be considered as such. But with that said these techniques will create a strong level of communications security.

Communications Security Considerations by Category

Sustainment level communications have the lowest COMSEC requirement. That said, they cannot be completely ignored. In the event of a disaster or even during routine communications there is the possibility of interception by those with ill intent. Further, transmissions must still be kept to the shortest amount of time possible to keep the net clear. Digital communications modes found in andFLmsg are used in some volunteer relief organizations due to the amount of data that can be transmitted in a quick amount of time.

Tactical level communications obviously have a far higher COMSEC requirement. The basic rule when using an analog radio is to run the lowest power possible to only communicate over the distance necessary and no further with limited antennas as described in the last chapter. That said, tactical communications being immediate in nature, they rarely have the time to encrypt and decrypt utilizing the methods described in this chapter. The focus on more sophisticated tactical radio systems is in addressing encryption built into the radio in real time. But, as noted, this creates a distinct pattern and signature, along with needing the equipment itself. The Guerrilla band must fight with what it has, and in most cases concerning communications that will be the Baofeng radio and little else. The trigrams as described later in this chapter is one of the older, but highly useful, methods for both obscuring the message as well as keeping the transmission time as short as possible.

Clandestine level communications have the highest COMSEC requirement by nature. These first encrypt the message before being either sent over analog means or via a digital burst. The encryption methods detailed in this chapter, Trigram and One Time Pad (OTP) cipher, provide the most robust means of encrypting a message. These can be used over literally any means, from written messages to those sent over the radio and anything in between. It must be stated, however, that the use of Trigrams and especially OTP will immediately attract the highest amount of attention from an agency tasked with counterintelligence; so while Trigrams may present strong encryption and OTP mathematically unbreakable, it will dedicate assets to locating the source.

Baofeng Digital Operation

Interfacing a Baofeng radio with a tablet or other mobile device for digital communications is relatively simple, inexpensive, and enables keyboard-to-keyboard digital data bursts from one data terminal to another utilizing a free app called andFLmsg. Your transmission time is cut down to seconds, far more efficient in both the signal quality versus voice but also in the handoff from one terminal to another, greatly mitigating the possibility of human error. This can also be done with the app's full version, FLdigi, on laptop computers. AndFLmsg, while stripped down, is far more user-friendly and simple to troubleshoot. This text will be using andFLmsg specifically.

The tools required are a Baofeng APRS cable, which inserts into the audio jack of the tablet and the two-prong microphone jack of the radio. Your radio must now be set up with the following steps to transmit the audio emitted from the tablet:

1. **Set the Radio to VFO mode, having both the Transmit and Receive frequencies on the display.**
2. **Ensure TDR (MENU #7) is set to ON.** This enables the radio, and the software, to receive on both frequencies simultaneously.
3. **Set the radio volume to 2/3 max.** This keeps the signals being received into the tablet's software from distorting the signal.
4. **Set the tablet volume to maximum.**
5. **Set SQUELCH (MENU #0) to 1.**
6. **Set VOX (MENU #4) to 1.** This enables the tablet's audio to automatically transmit over the radio. You will be controlling the transmitter through the software itself.
7. **Cut the Time Out Timer (MENU #9) to OFF.**

The Baofeng radio is now ready to be used with digital transmissions.

Using andFLmsg

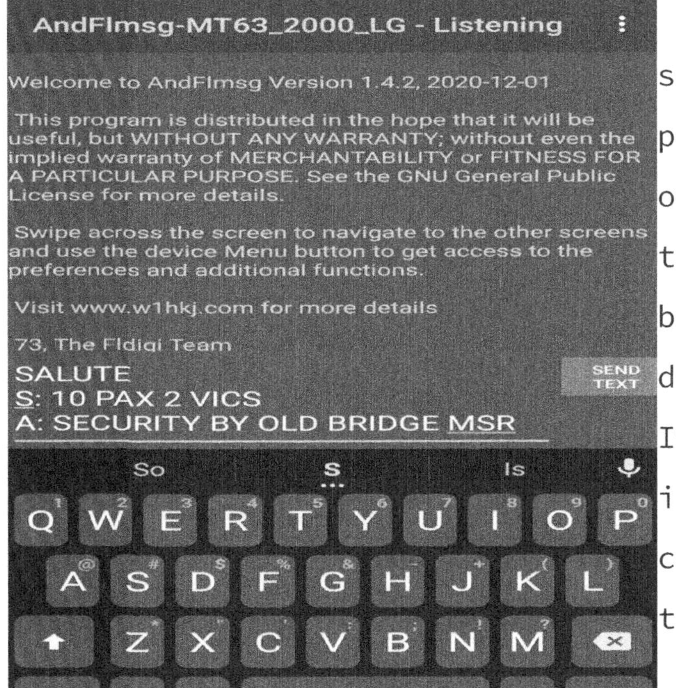

AndFLmsg is a free, open source digital data protocol program with 8 different overarching protocols, each with their sub-modes broken down by bandwidth, giving a total of 108 different communications modes. It is extremely user friendly and intuitive, making digital communications simple. It has three unique screens which can be indexed by swiping the screen of the tablet. The first screen is the terminal to enter quick messages.

When opening the program, you'll note at the top the operating mode. In the picture it is set to MT63 2000 LG, or the broadest bandwidth version of MT63 with a long interleave, meaning a slightly longer transmission but more redundant data being sent for forward error correction. This is one of the preferred operating modes in the field. Once the report is complete, hit SEND TEXT and the program will begin to transmit the digital encoding. The mode, in the top bar, will change from blue to gold while transmitting and back to blue when finished.

You can change the operating mode two ways: by clicking the three circles in the top right corner of the screen and selecting SETTINGS, then selecting CUSTOM LIST OF MODES, then checking the boxes of the operating modes to which you want to limit the device. The other way is to swipe to the second screen, the receiving terminal, and hit NEXT MODE or PREVIOUS MODE.

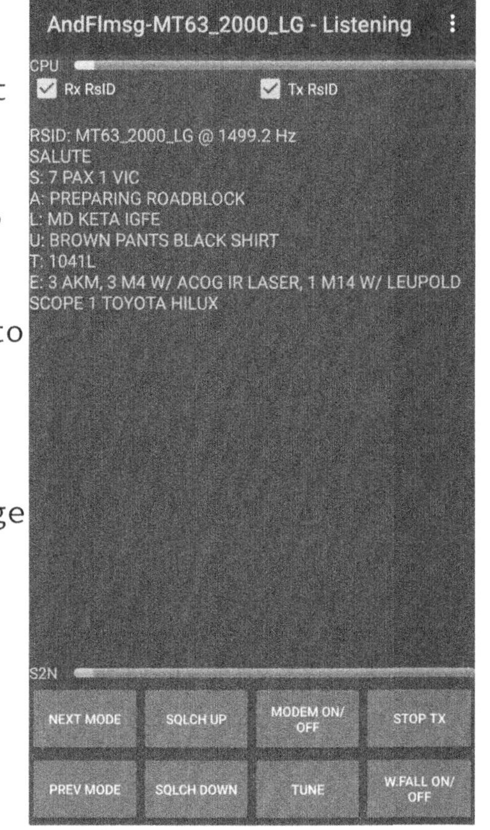

Once you've sent your message swipe to the second screen. This is where received messages will be displayed. Note in the picture the terminal has received a message in MT63 2000 LG with the message header clearly stating the nature of the traffic (SALUTE) and the lines of the message clearly laid out. The total transmission time is around 3 seconds, whereas this message would have taken a significantly longer time to send via voice.

Note the buttons at the bottom. NEXT and PREV MODE controls changing the modes. SQLCH UP and DOWN control the level of squelch. ONLY HAVE THE SQUELCH AS HIGH AS NECESSARY. You can visually see the squelch level in the S2N (signal to noise) bar just above the buttons. If the squelch is too low, random characters will fill the terminal - this is called digital static. If it is up too high, you may miss messages. As you are receiving messages the S2N will fill up - this is a visual indicator of incoming traffic.

AndFLmsg also has the ability to send pictures, files, and dedicated report forms. These are importable / exportable as .html files, making it simple to create your own pre-made report formats that can be filled in as needed.

To send pictures, swipe to the third screen and hit COMPOSE. This will bring up all of the options for forms. Scroll down to PICTURE and long press it to bring up the blank picture form. Press ATTACH PICTURE to attach pictures from the device's memory. Once you've completed this, hit SAVE TO OUTBOX at the bottom left corner of the form and RETURN at the top to go back to the previous menu. Now hit OUTBOX to

open the outbox, and your message will be there. Long press to open it up.

If everything looks correct, you're ready to send the message. To do this, first note the modes. There are two displayed - MT63 2000 LG and MFSK64, along with the amount of time the program has calculated it will take to send the message. The program will be using these two modes. MFSK64 is an extremely fast, wideband mode to send images and is the default setting. While there are other modes capable of sending images, they are very slow in comparison. Once you are ready to transmit, hit TX OVER RADIO and the message will begin transmitting.

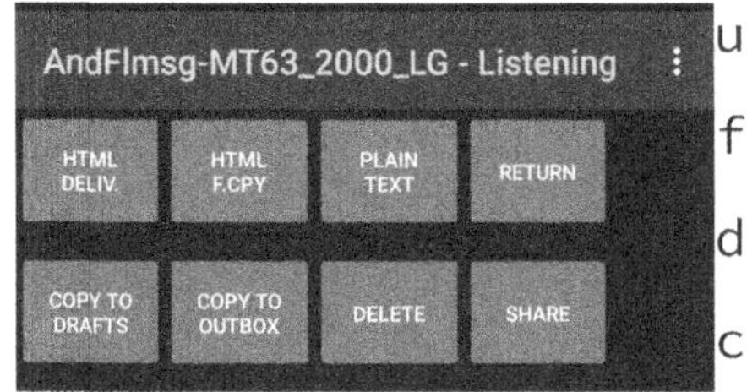

To open received message forms, we go back to the terminal and press INBOX. There will be a new message, and just like before, we long press to open up the contents.

Trigram Encryption

One of the common requirements of radio, regardless of analog only or digital operation, is the need to be as brief as possible on the air. This is not just a COMSEC requirement but also to keep transmissions efficient.

Trigrams themselves are a means to both shorten and encrypt a message. It takes three letters in substitution for a word. A sample list of these are found in Appendix C and will be used as the example in this manual. Amatuer Radio operators do this utilizing what's known as Q codes, not for encryption but to shorten the overall message. This owes to the original conception of the trigrams being sent over CW, or Morse Code.

Back in the early days of telegraphs and then the advent of radio, the primary means of sending traffic was via Morse Code. Messages were made short by a standardized series of trigrams. This skill continues today with the aforementioned Q codes, but in the clandestine world those trigrams are randomized, requiring the current key to decrypt the message. Otherwise, anyone intercepting it is simply getting random characters.

Covert teams in Europe, South America and Asia during the Cold War used trigrams extensively, tapping out the Morse Code letter groups on a tape recorder then speeding the device up to burst the transmission over an analog radio. Lawrence Myers would write about this technique in his book *SpyComm* as a method frequently used over CB for coordinating local clandestine cells in South American countries he operated in for the CIA. The same technique can be used today with andFLmsg and a Baofeng.

Generating the Trigram List

What I suggest is entering the entirety of the list in a spreadsheet in four columns on an air gapped computer. The first three columns have single letters and the fourth the words themselves from the master list. Set the columns to shift at different intervals, and you've created a strong method of encryption that first must be identified by a cryptologist then ran through a data set to break. It is possible, but extremely time consuming on the part of an adversary. By the time they decrypt the message, assuming they've intercepted the message and recognized the digital mode in which it was sent, the action itself has likely already taken place. It becomes useless.

Using Trigrams

Encoding messages via a trigram list is a simple word substitution. The words in the list are organized in an ENCODE and DECODE list. The ENCODE list has the words alphabetized, the DECODE list has the trigrams alphabetized. This makes both functions quick with practice.

Let's take a look at a sample message. You're reporting that fighters have been killed in an ambush from a concealed group of paratrooper infantry in the area. Your message might look like this:

FIGHTERS KILLED IN ACT APPROACHING CAMOUFLAGED PARATROOPER(S) INFANTRY

This is a quick message to encrypt from our sample list:

FIGHTERS (BQL) KILLED IN ACT (WIO) APPROACHING (BKF) CAMOUFLAGED (JNU) PARATROOPER (OAM) INFANTRY (BLT).

As a method of keeping messages simple, I always terminate the message with ZZZ to indicate nothing follows. This prevents any confusion. Your message becomes:

BQL WIO BKF JNU OAM BLT ZZZ

Transmitting this is much faster and far more secure than saying it in the open, and when sent via a data burst, reduces the transmission time to just over a couple of seconds. While it is possible that the message can be intercepted, the likelihood, when utilizing the other methods for transmitting described in this manual, are slim.

One Time Pad

One Time Pad (OTP) encryption is also known as a Vernam Cipher and is, when used a single time, mathematically impossible to break. OTP functions first by utilizing a codex to assign numeric values to each letter of the alphabet. It then takes a random series of numbers, broken down into groups of five digits each, and aligns them into a block.

A-1	B-70	P-80	FIG-90
E-2	C-71	Q-81	(.)-91
I-3	D-72	R-82	(:)-92
N-4	F-73	S-83	(')-93
O-5	G-74	U-84	(,)-94
T-6	H-75	V-85	(+)-95
	J-76	W-86	(-)-96
	K-77	X-87	(=)-97
	L-78	Y-88	(?)-98
	M-79	Z-89	BRK-99
			SPC-0

32244

32244 52687 97412 86319 11011 59341 73741 29248 65123 56878

19652 15821 01512 44462 36799 12491 08122 13421 79512 72121

06503 41072 61772 22229 13714 24763 19975 59821 63483 64465

14324 86922 24922 62231 52134 70252 14319 23219 92012 08194

11119 21919 67421 01254 25722 42491 16132 71151 86117 01825

69138 19415 21792 46102 29541 17221 16413 65714 91978 62353

42551 58181 59165 14216 21238 87777 42524 61411 92206 11872

22129 20919 51252 20934 85440 22922 21922 06210 99213 32819

51261 86248 15863 14618 52148 91232 24524 14388 71341 92230

The OTP example here has the codex above and then sample key. The codex normally remains the same across all of the messages. The OTP key is the component used a single time. The key is marked by the numbers **32244** the very top and is repeated as the first five digit block. There are two copies of this key; one for the transmitter and one for the recipient. This is the identifier, telling the recipient which key to use to decrypt the message and will not be encrypted with letters.

The first step in encryption is to take our message and assign the individual letters of the message their corresponding number values from the codex. Based on the example above, the sample message 'Meet at the old church at five PM Tues' would be encoded in numbers as:

M e e t _ a t _ t h e _ o l d _c h u r c h _a t_ f i v e_p m_ t u e s

79 2 2 6 0 1 6 0 6 75 2 0 5 78 72 0 71 75 85 82 71 75 0 1 6 0 73 3 85 2 0 80 79 0 6 84 2 83 0

These are then broken into five character groups. The spacing itself is irrelevant, but is done to keep the math simple. OTPs encrypt and decrypt based on simple math:

To encrypt, we subtract the numbers in the Key from the encoded message, one digit at a time. The sum becomes the transmitted message. To decrypt, we place the received message on top and the key underneath, and add one digit at a time. This returns us to our codex numbers which are then decoded into letters.

The message below displays the message encoded via the codex on Line 1. The OTP key is listed on Line 2, and the encrypted message is contained on Line 3:

Line 1: - - - - - 79226 01606 75205 78720 71758 58271 75016 07338 52080 79068 42830

Line 2: 32244 52687 97412 86319 11011 59341 73741 29248 65123 56878 19652 15821

Line 3: 32244 27649 14294 99996 67719 22417 85530 56878 42215 06212 60416 37019

NOTE: if the bottom number is greater than the top number, add 10 to the top number. When decrypting the sum will be greater than 10, and you simply drop the 1 leaving the second digit remaining. For example, the last digit in the first block is 6, and the key has a 7. The 6 becomes 16 and the sum, in the transmitted message, becomes 9. When decrypting, we'd add 9 and 7 to make 16, dropping the 1 to leave 6.

Decrypting is the same process in reverse. We received the transmitted message and now line up the numbers from the key underneath them to begin adding a single digit at a time. This, when done properly, will give us the numbers from our codex that we can then use to decode the individual letters.

Line 1: 32244 27649 14294 99996 67719 22417 85530 56878 42215 06212 60416 37019

Line 2: 32244 52687 97412 86319 11011 59341 73741 29248 65123 56878 19652 15821

Line 3: - - - - - 79226 01606 75205 78720 71758 58271 75016 07338 52080 79068 42830

The message then becomes:

79 2 2 6 0 1 6 0 6 75 2 0 5 78 72 0 71 75 85 82 71 75 0 1 6 0 73 3 85 2 0 80 79 0 6 84 2 83 0

M e e t _ a t _ t h e _ o l d _ c h u r c h _ a t _ f i v e _ p m _ t u e s _

Notice the math worked out. I always, as a rule, encrypt, then decrypt my messages before sending them to verify the accuracy and prevent any confusion. Further, the use of grid paper helps tremendously in keeping the digits aligned. The most common mistakes made in encrypting and decrypting OTP is by misaligning the number columns. One digit can throw the entire message off. Keep it one digit at a time, working methodically, and you'll be good to go.

As a COMSEC and Operational Security (OPSEC) concern, the remainder of that pad, regardless of the fact we did not use many of the digits, cannot be used again. **The OTP is only truly random when used once**. Constructing the pads themselves can be done a number of ways. One method is by using 5 10 sided dice to generate the random numbers for the pad. Another is to obtain a commercially available OTP printer called an ADL-1.

Truly Unbreakable

OTPs are mathematically unbreakable when used one time, presenting the highest level of security possible. Brevity remains the watchword, however, and the method I strongly advocate is mating the two concepts for encryption described here, the use of Trigrams and OTP, together to create a shortened burst that is truly unbreakable. It does however take time and skill, but for the most robust security possible when sent by virtually any means, it is unbreakable by anyone but those with both keys. Two layers of security presents an advantage to an underground organization that even if one key is compromised, the other may not be. For a Guerrilla force operating in the shadows, every precaution must be made regarding COMSEC. Utilizing data bursts, coupled with the layers of encryption described here, creates the most robust form of COMSEC over the air.

APPENDIX A: THE BAOFENG REPEATER

Disclaimer This design is for a cross band repeater. Before building this please understand how the FCC rules apply.

This tactical/portable repeater design will allow for a UHF/VHF cross band repeater operation using cheap and readily available COTS products. The sealed box, solar panels, and battery will, depending upon use, ensure extended operation that could potentially run for years. Although any radios could be used, this design will utilize Baofeng UV-5R HT radios. For an optional upgrade, get an 8-watt UV-5R8W HT on the Tx side of the repeater pair. A sketch of the overall design looks something like this:

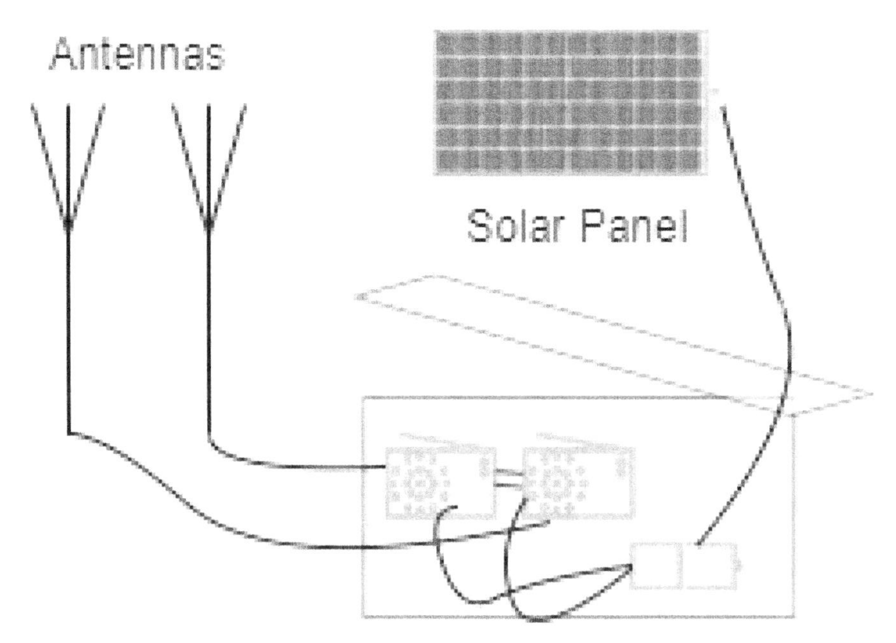
Antennas
Solar Panel
Ammo Can (2 radios & battery)

Ammo Can Setup

Parts List

- 30 Cal ammo can

A typical 30 Cal ammo can is the perfect size for this project. To allow for antenna and solar panel cables, a 1" hole is drilled into the ammo can. Use a 1" clamp connector (obtained from any hardware store) to protect and secure the wires. Silicone can be applied to waterproof the box.

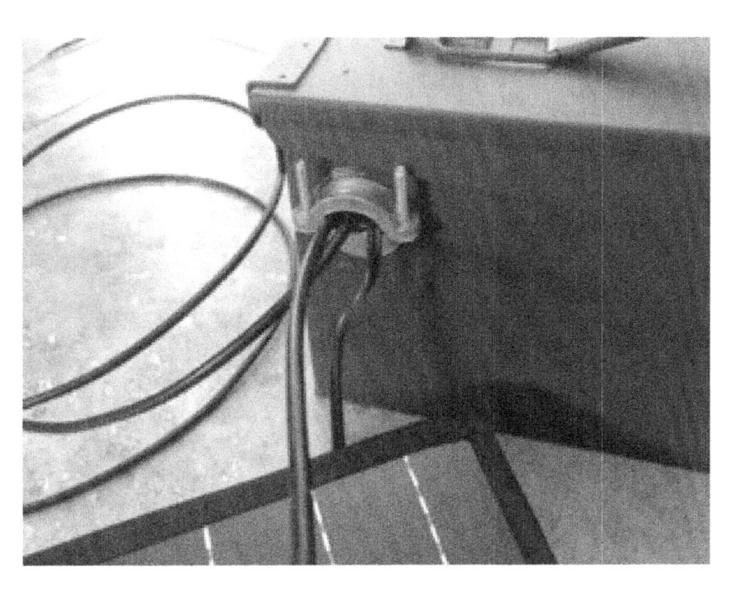

These are additional pictures of the interior of the can as well as a completed project picture:

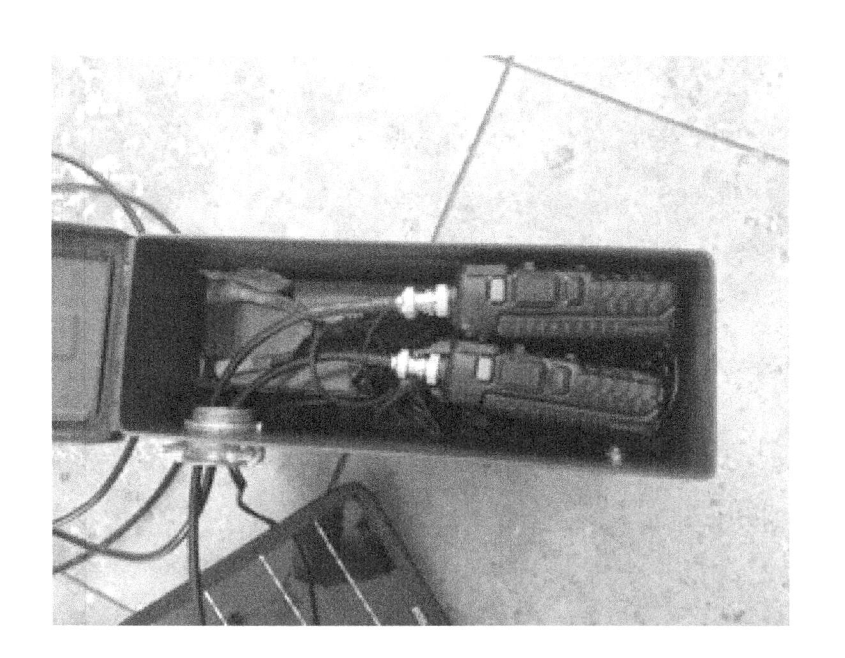

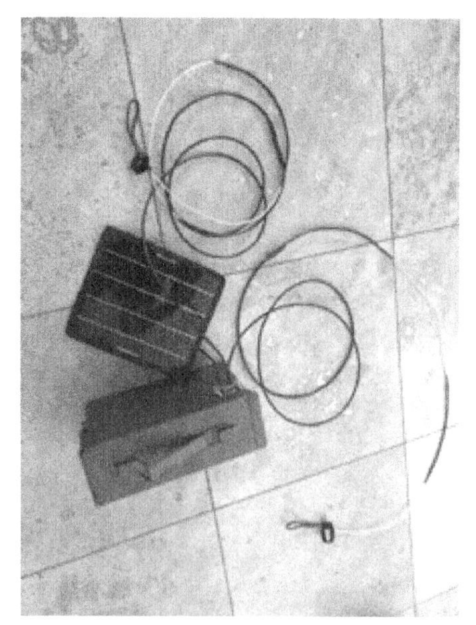

Power Setup

(Note: This part is optional but is an important part for extended operation.)

** Parts List **

- standard 7.4 Ah lead acid battery
- Cigarette ‘Y’ adapter
- 5-Watt Solar Panel
- SAE extension cable
- 2 X UV-5R Car charger adapter
- 10-12awg Female Spade Crimp

Leaving enough wire to easily work with, cut off the male end of the ‘Y’ adapter and cut the SAE cable in half. Strip both wires on the end of the ‘Y’ adapter and the SAE cable. Twist the red wires together and attach crimp the spade adapter. Do the same thing for the 2 black wires. Connect the system as shown in the pictures below:

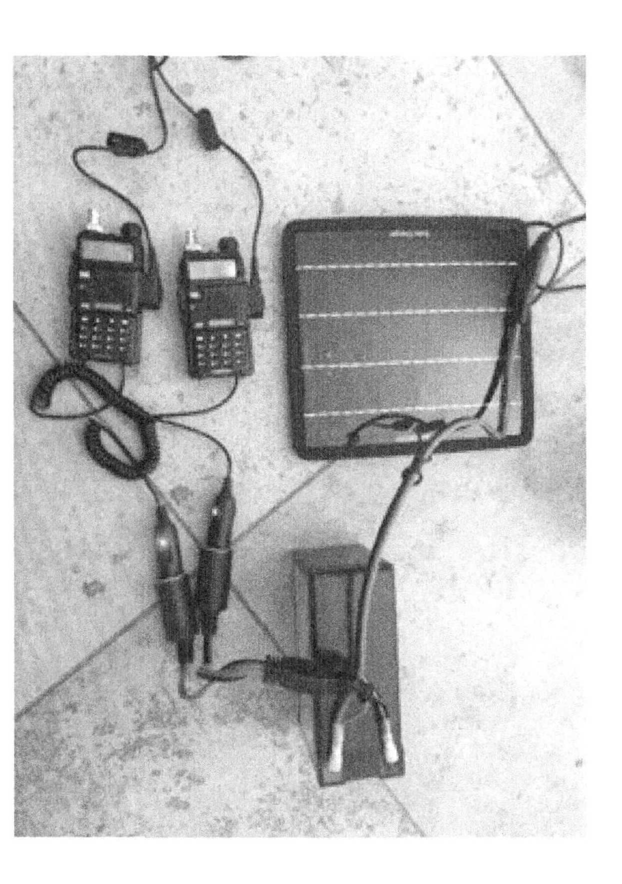

If you’re looking for the simplest design, you could skip the ammo can, battery, and solar panel and just go with something like this:

Repeater Setup

** Parts List **

- 2 X Baofeng UV-5R
- Baofeng UV-5R8W
- Repeater box (Note: I have had good success with the ‘YiNi Tone RC-108 - I can no longer find this part on amazon. I have, however, had success finding this part at Walmart.com and on ebay.com - Search for ‘RC-108 Repeater Box’. There are other options on amazon, but I have not yet tested them)

This guide covers programming these radios from the face, but also includes screen shots from Chirp in case you would rather program them using a programming cable. See Appendix B for general CHIRP settings that is recommended for all the radios.

In either case, before you start programming the repeater pair, you will need to determine 3 things:

- Transmit (Tx) frequency. (In our example we are using 146 MHz)
- Receive (Rx) frequency. (In our example we are using 447 MHz)
- CTCS Tone. (In our example we are using 67 Hz)
    - Although it doesn't really provide security, having a R-CTCS tone set on the Rx side of the repeater pair does help prevent unwanted users from activating the repeater.

(Note - any frequency that your radios support can be used by this repeater)

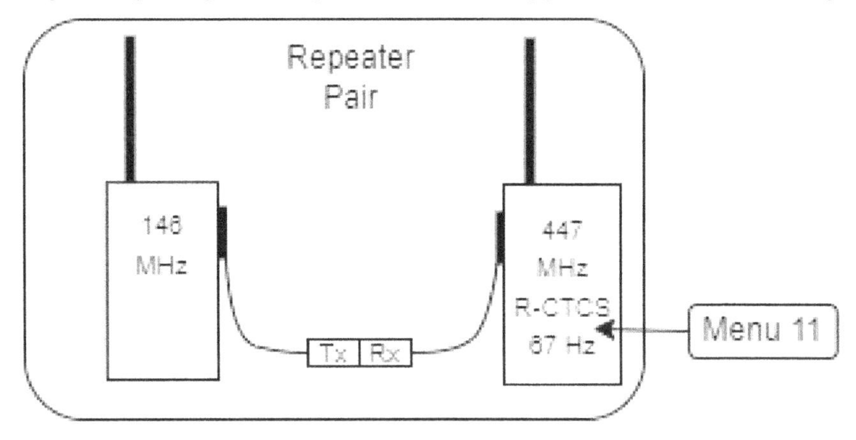

Repeater Receive (Rx) programming

Radio face method:

1) Enter VFO mode
2) Type 447.000 (or whatever other frequency you desire)
3) [Menu] 11 [Menu] 67 [Menu] [Exit]
   a. (Note: instead of typing ‘67’, you can also press the up/down arrows until you see your desired tone frequency)

CHIRP method:

Settings	Loc	Frequency	Name	Tone Mode	Tone	ToneSql	DTCS Code	DTCS Rx Code	DTCS Pol	Cross Mode	Duplex	Offset	Mode	Power	Skip
	1	447.000000	RX HIGH	TSQL		67.0					(None)		FM	High	

Repeater Transmit (Tx) programming

Radio face method:

1) Enter VFO mode
2) Type 146.000 (or whatever other frequency you desire)

CHIRP method:

Settings	Loc	Frequency	Name	Tone Mode	Tone	ToneSql	DTCS Code	DTCS Rx Code	DTCS Pol	Cross Mode	Duplex	Offset	Mode	Power	Skip
	1	146.000000	TX LOW	(None)							(None)		FM	Low	
	2	146.000000	TX HIGH	(None)							(None)		FM	High	

Once your radios are programmed, plug the repeater box into the radios. Make sure you plug the Tx side of the box into your Tx repeater and the Rx side of the box into your Rx repeater.

Turn the radios on – I'm not sure if it matters much, but I have had good success turning the volume up 50%-75%.

Team Radio Setup

Before we can begin programming our team radios, we'll need to figure out our Offset. We do this by subtracting our Rx frequency from our Tx frequency. If the number is positive, we will have a positive offset. If it is negative, we'll have a negative offset. In our example we subtract 146 from 447. This is a positive offset of 301.

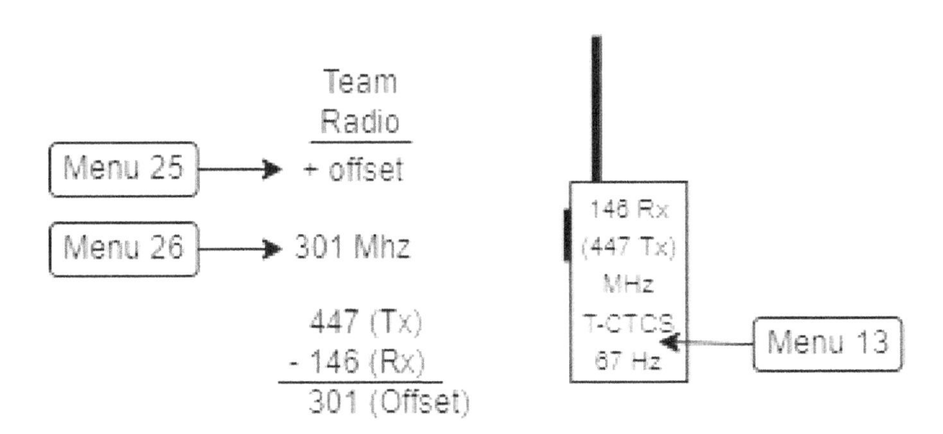

Radio face method:

- Enter VFO mode
- Type 146.000 (or whatever other frequency you want - make sure and figure out the proper offset)
- [Menu] 25 [Menu] <press up/down arrows until you get '+'> [Menu] [Exit]
- [Menu] 26 [Menu] 301000 [Menu] [Exit]
- [Menu] 13 [Menu] 67 [Menu] [Exit]
    - (Note: instead of typing '67', you can also press the up/down arrows until you see your desired tone frequency)

CHIRP method:

Settings	Loc	Frequency	Name	Tone Mode	Tone	ToneSql	DTCS Code	DTCS Rx Code	DTCS Pol	Cross Mode	Duplex	Offset	Mode	Power	Skip
	1	146.000000	RPTR L	Tone	67.0						+	301.000000	FM	Low	
	2	146.000000	RPTR H	Tone	67.0						+	301.000000	FM	High	

Antenna Build

The antenna build is optional - you will need an external antenna, however, if you plan on using an ammo box to contain your repeater. If you do not want to build antennas, there are good options on ebay. One type that is good for this application is the 'Dual band Slim Jim' or the Jungle Antenna. Just make sure you get the proper cable end for you radio. I prefer to get a BNC adapter and use antennas with a BNC end.

** Parts List **

- SMA-F to BNC-F adapter (2 pack)
- 20' RG-58 coax cable
- Heat shrink tubing (various sizes)
- BNC-M crimp connector

When building this antenna, make sure to measure very carefully. With my 1st few attempts, I was off by 3-5mm and had a very large SWR. As I became more precise, my SWR @ 145 MHz was around 1.2

You will need a soldering iron, wire cutters, coax strippers, a box cutter or razor blade, and a crimper for this project. I used heat shrink tubing on the outer insulators to keep them from moving and heat shrunk a rope loop on the end of the antenna as an attachment point. I also soldered about 5’ of antenna feed line directly to the antenna. I then crimped a BNC-M connector on the end. Diagram and photos are below:

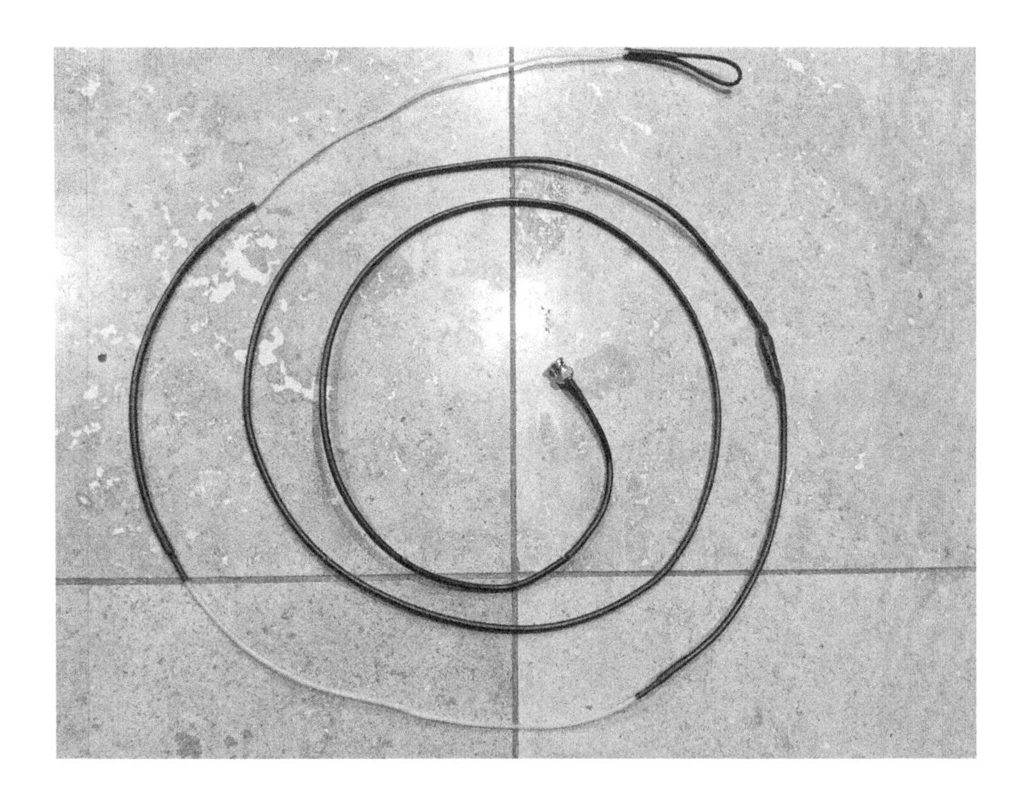

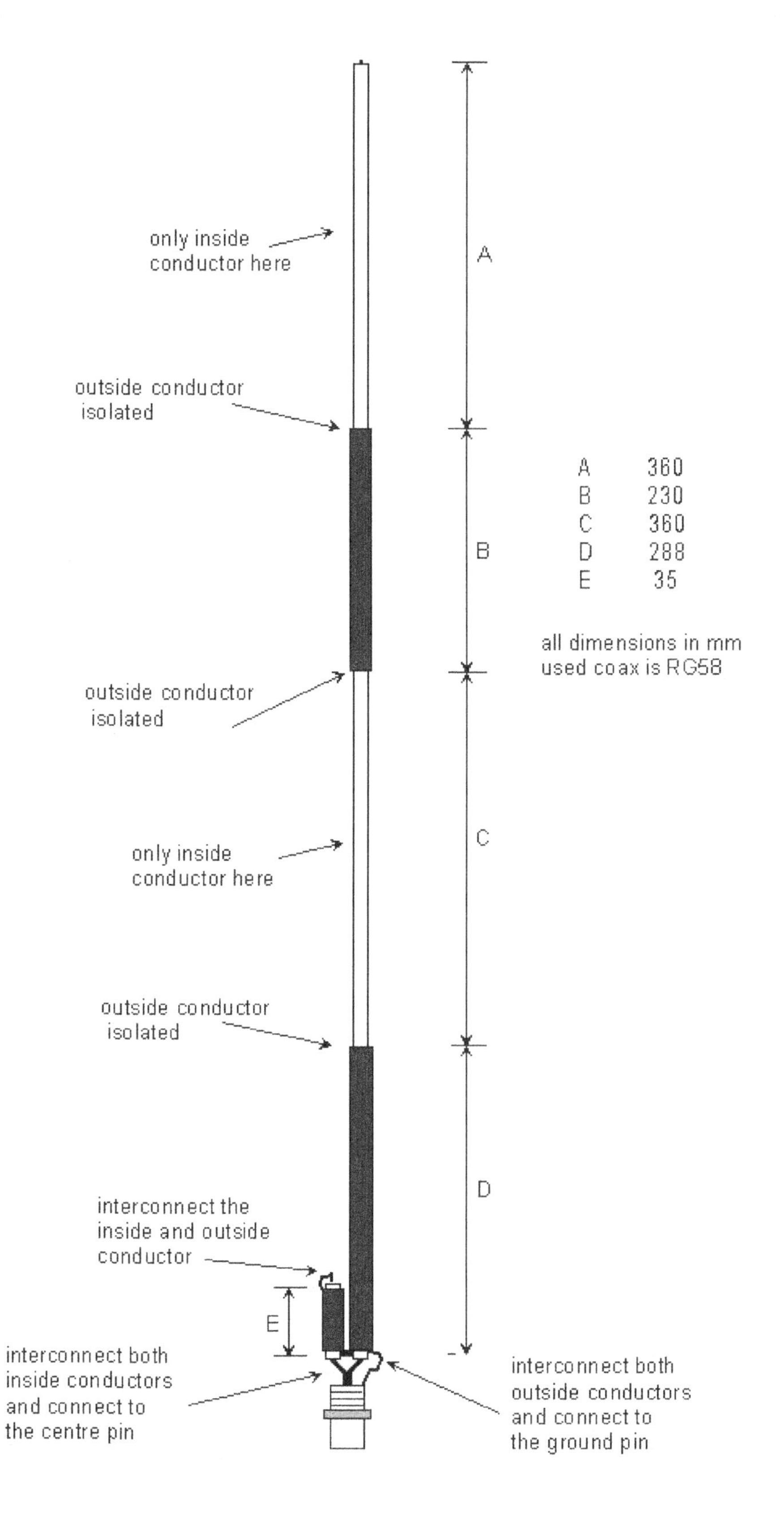
only inside
conductor here
A
outside conductor
isolated
B
A 360
B 230
C 360
D 288
E 35
all dimensions in mm
used coax is RG58
outside conductor
isolated
C
only inside
conductor here
outside conductor
isolated
D
interconnect the
inside and outside
conductor
E
interconnect both
inside conductors
and connect to
the centre pin
interconnect both
outside conductors
and connect to
the ground pin

APPENDIX B: SAMPLE SOI AND REPORT FORMATS

Signals Operating Instructions (SOI)

Primary [FREQ / MODE]:
PROWORD:
TX:
RX:
Alt. Freq [FREQ / MODE]:
PROWORD:
TX:
RX:
Contingency Freq:
Emergency Signal:

Callsigns:
TOC / Control:
Element 1:
Element 2:
Support/Recovery:
Challenge / Password:
Running Password [DURESS]:
Number Combination:
Search and Rescue Numerical Encryption Grid [SARNEG]:

K	I	N	G	F	A	T	H	E	R
0	1	2	3	4	5	6	7	8	9

0	1	2	3	4	5	6	7	8	9

*** SARNEG first number rotates with last numeral of previous date***

COMMO WINDOW SCHEDULE:

	1	2	3	4	5	6	7	8	9	10
• DAY:										
• NIGHT:										

Tactical / Immediate Intelligence Reporting

SALUTE [PRIMARY]
Size
Activity
Location
Uniform
Time
Equipment

SALT [SUPPLEMENTAL]
Size
Activity
Location
Time

MEDEVAC Request

LINE 1: LOCATION OF PICKUP SITE
LINE 2: CALLSIGN / FREQUENCY
LINE 3: NUMBER OF PATIENTS BY PRECEDENCE
- # URGENT (IMMEDIATELY PUSHED TO SURGERY)
- # PRIORITY (WILL SURVIVE 12-24 HRS)
- # ROUTINE (INJURIES NOT LIFE-THREATENING)
- # EXPECTED (DOA)

LINE 4: SPECIAL EQUIPMENT REQUIRED
LINE 5: NUMBER OF PATIENTS
- # LITTER - BORNE:
- # AMBULATORY:

LINE 6: SECURITY OF PICKUP SITE
- N: NO ENEMY TROOPS
- P: POSSIBLE ENEMY IN AREA
- E: ENEMY PRESENCE
- X: ACTIVE ENGAGEMENT / ARMED ESCORT

LINE 7: METHOD OF MARKING CASUALTY COLLECTION POINT / PICKUP SITE
- A: PANNELS
- B: PYRO
- C: SMOKE / COLOR:
- D: NONE / OTHER METHOD

LINE 8: PATIENT STATUS
- A: YOUR COMBATANTS
- B: ALLIGNED CIVILIANS
- C: FRIENDLY / ALLIED ARMED GROUP MEMBERS
- D: UNALIGNED CIVILIANS
- E: ENEMY WOUNDED

Strategic Cabal / Report Formats

- **MESSAGE HEADER:** ANSWERING STATION, THIS IS CALLING STATION [CALLSIGN DE CALLSIGN] // MESSAGE NUMBER // MESSAGE PROWORD // DURESS CODEWORD
- **MESSAGE FOOTER:** END OF MESSAGE [EOM] // ACKNOWLEDGE [ACK] // INITIALS OF RTO

ANGUS: Initial Entry
AAA. Date / Time / Group
BBB. Team Status
CCC. Location
DDD. Deviations
EEE. Additional Information

CYRIL: Situation Report
AAA. Date / Time / Group
BBB. Current Loation
CCC. Medical Status
DDD. Equipment Status
EEE. Supply Status [Batteries, Ammo, Water, Food]
FFF. Team Activity since last Commo Window
GGG. Team Activity until next Commo Window
HHH. Remarks

BORIS: Intelligence Report
AAA. Date / Time / Group
BBB. DTG of Observed Activity
CCC. Location of Observed Activity
DDD. Observed Activity
EEE. Description of Personnel, Equipment, Vehicles, Weapons
FFF. Team Assessment

CRACK: Battle Damage Assessment
AAA. DTG
BBB. Type of Target
CCC. Description of Target (Physical and Functional Damage)
DDD. BDA Analysis

UNDER: Cache Report
AAA. DTG
BBB. Type
CCC. Contents
DDD. Location
EEE. Depth
FFF. Additional Info / Reference Points

APPENDIX C: SAMPLE TRIGRAM LIST

ENCODE LIST:

RSU (SPACE / NULL)
LIH a
LSY abdomen
CEQ able/ability
AFK abort-s-ed-ing

WYY about
WNC above
ONA accident-al-s
AVF accomplish-ed
KYH accomplishing

QAH ache
MAY acknowledge-d-ment
XWF act-ion-ing-s
PVC act-ive-ivity-s
PSA add-ed-itional

DBO add-ing-s
EMI adjacent
CSH administration
CCD administrative
GVA admit

KFE advance-d-ing
XUF after
RUS age
HLV agent-s
IEU agree

RBR air (fld/port)
JBI airborne DF platform
BPS aircraft
BFF airdrop-s
NSM airmobile

TRM alarm
WRM ALE
PCQ alert-ed-ing
VJU alive
JWS all

LWF approach-ing
BNJ approve-al-d-ing
PMT approximate-ly
ADY April
XFY are/is

QEP area-s
APK arm-s-ed
PGB army-ies
XVY around
QNT arrange-d-ing

HSN arrange-ment
CIL arrive-al-d-ing
GOU arrow
PQA article
AAS as

JML asap
BPX ask
LID assault
UAR assemble-d-ing
CCI assist-ance

SQG assist-ed-ing
QMD at
HME at (once)
CJA atlantic
CCY attach-ed-ing

IQN attach-ment
QNN attack-ed-ng-s
JKC August
DUM authorization
MAV authorize-d-ng

NUR automatic
YYL automobile
QNS available
DMR aviator/pilot
TBO axe

IFW bed
WVE been
MHN beer
JVM before
RPU begin

UCK below
SBO belt
GXU between
YYY beyond
BSV bike

RWI binoculars
XKJ biological
FRS bit
ADD bite
HHC bivouac-ed-ing

YRX black
UMF bleed
CGI blister
ARH block-ed-ing
SIX blow-en

OTL blow-ing (up)
GHD blue
YDL boat-s
IAE bomb-d-r-ng-s
RIL bone

BVF boobytrap
OWI book
EDT boots
AVL border
FDV boundary-ies

THX bow
TLH box
LDH boy
BTB break/broken
PDU breakfast

CFQ alley
KGD alt supply pt
ENU alternate
VLU alternate commo
AJQ altitude

XTG am
AJS am in pos (to)
VEQ amateur
BPP ambulance
UVJ ambush-ed-ing

LII ammunition
VVI amplitude modulation
HTP amputate
EOF an
LNJ and

WFT answer-d-ing-s
QHB antenna
BTJ anti-
RGL apply
BKF approach-ed-s

VDW can
MDT canal-s
BUJ cancel-led-s
QYA candle
SAP capable-ity

IPF capacity-ies
JIA capsule
CDR capture-d-ing
CBR car
OFP carry-ied-iers

YOR casualty-ies
VPN cave
JKE CB radio
GPV cease-s-d-ing
RWH challenge-d-ing

GDM chance
RQP change
FLB change to/from ciphe
QMS channel

VXT axis (of)
UKR azimuth (of)
DJE b
GPF back-ed-ing
BNP bad

VTN ball
KGR band
VNS bank
VEA barbed wire
FBN barometer

AME barrage-s
UBL barrier
TUQ base-d-s-ing
BDB battalion-s
RWA battery charger

ATR battery-ies
TON battle-s
NLF baud
VNB be-en-ing
MXI beach-ed-s-ing

QXH concentrate-d-ing
JTA concentrate-s-ion
VGV condition-d-ing
RBP conduct-ed-ing
ASH confidential

ENO connect-ed-ing
CTX connect-ion
FWJ conservation
ACT conserve-d
KOO consolidate-d

WLY consolidate-ng
TTM consolidation
PKE contact-ed-s
MEQ contact-ing
OFX contaminate-d

NKD contaminate-ion-ing
NCM continue-ation
ITD continue-d-ing
NHI control-d-ing-s

PTK breast
OPK bridge-d-ing-s
AYA brother
RBT bug(s)
XME building-s

FRE bulldozer
UVC bunker-s
DAW burst-ing-s
WRT but
CDA by

YJM bypass-d-ing-s
AFL c
XBG cable-s
MDA cache'
KTD caliber

KHV call
UCV camera
JNU camouflage-d
XYV camouflage-ing
YBS camp

BOC decontaminate-d-ion
CEV decrease-d-ing
LAO deer
JKT defend-ed-g-s
IJW defense-sive

WVH degree-s
AUQ delay-ed-ing-s
WVK delete-d-ion
GBT deliver-ed-ing
YIC demolition

YMQ demonstrate-d-ing
SPR demonstrate-d-s-ion
QKQ deny-ied-ies
PHF depart-d-g-ure
DRR deploy-ed-ing

QTF deploy-ment-s
CQQ depot
WYC deputy-ies
XAD designate-d

PYE charge-d-s

EIL check-d-ing-s
QTQ checkpoint
JJK chemical-s
NON child/children
FPN church

UNU circle
BLY city
RTP civilian-s
WFC clarification(of)
JLY class-ify

TKW classification
MQE clean-ed-ing
AEX clear-ed-ance
BMA clearing-s
TAT close-d-ing

NLJ cloth-es-ing
LNA coast-s-al
YXG coffee
KEM collapse
AKO collect-ed-ion

TAB collect-ing-s
NKX column-s
EII combat
GWA come-ing
JKO comma

KCU command-er-ed-ing-s
CPC commit-ment
NHT commit-ted-ing
LAG communicate-d-ions-s
CJL company

EDW complete-d-ing
PKN complete-ion
GQK compound
VKC compromise-d-ing
EKY computer

XSX due
VHM dump-ed-ing
NYL during

UFW conversation

DVX convoy-s
KKO cook
OHB coord-s-d-ing
EQG corporation
UGC corps

YGA correct (me)
UWU correct-ed-ing
APN cost
LNN could
CTW counterattack-ed-s

HHX counterattack-ing
YLI counterpart
UHM cover-ed-ing
GHK craft
BFH crane

IXN critical-ly
LWY cross-ed-ing
CDL crossroad-s
XYY crushed
OYY crystal

TOF CTCSS
GQN culvert
CMG cup
IWG cut-off
KPM CW

NRY d
KMB damage-d-ing
VKL danger
BUK dangerous
VDJ darkness

RVX day-s/date-s
CVV deadline-d-ing
PXF death
QTB December
MFL deception

PIN expose-d-ing
CUA extend-d-g-ive
AYH extend-extent

CAU designate-ion

RNI desk
KDQ destination
NOH destroy-d-ing-s
LEP detach-ed-es
BPV detach-ment-s

MTN determine
ONT diesel
IFU dig/dug (in)
PJK dinner
BFR direct-d-on-ng

OLV direction finding
QOW disable-d-ing
SBE disapprove-d-al
PQJ disperse-d-g-l
SMN displace-d-ing

WNW displace-ment
CUW dispose-d-al
GGW disregard-d-ing
ISL distance-s
LJK district-s

EDC ditch
LAD divert-ed-ion
QSF divide
SVW division-s
RCL do

NFC dog
RUN down-ed
NCF downed acft
UGJ drill
KIE drip

CNK drive
YVP driveway
WDX drop-ping-ped
FBP drug-s
MDV DTMF

ERM frequency-ies
KKG Friday
GMS friend-ly

XVE e
LWB e-mail

MYY east-ern (of)
PYN ed
YQC effect-ed-ing
HDF effect-ive-s
CPL eight

LRB electric-al-ity
UUL element-s
KCQ elevate-d-ion
QPW eliminate-ed-ing-s
LFC elk

FHG embank-ment
BDI embrace-d-ing
EVE emergency
UCB emplace-ment
PWE employ-ed-ing

HFM en
QEY en-ed-ing
FEQ encounter-d-ing
PDV encrypt-ed-tion
DBP enemy

IIB engine
LDV engineer-d-ing-s
ARV enlist-ed-ing
VHP entrench-ed
SOF envelop-ed-ing

PRA envelope
NDM equip-ing-s
BWD er
WGE error
QAO es

BPJ escape-d-ing
VSD estab-ed contact at
YED establish-d-ng
UVO estimate-d-ing
AGD estimate-ion

ABO et
TSG eta your loc

OTF extend-sion
UHG extra

NYT extract-s-d-ing
JYE f
OMU facility
AHW fade-ed-ing
DKU fail-ed-ing

SVI fallout
VWE family
AKP far
UPQ fatal
NLN father

RKF favor-able-ing
JBY favor-s-ed
VEV February
KSJ feet/foot
SPG FeldHell

DUJ FEMA
OFT female
PAO field-s
BQL fight-er-ing
BPI final

FNU fire-d-ing-s
UKM fish
GML five
RNT flame-d-ing-s
OYT flank-ed-ing-s

QKT flare-d-s
XVI fled
NQM flesh
RIF flexible-ity
UTF flight

ANG foe
BCY fog
MOS fold-ing
XQQ follow-ed-ing
XIV food

FFI for
MRS force-s-d-ing

MTV from
SEF front-age-ally

ETL FRS
AAR fruit
MHQ fuel
IGQ fuse-d
MNM future

WXW g
QRC gallon-s
NML gas-ing-ed
ICX gasoline
CND gate

LFJ general-ize
WYV generator(set)
PMV get-ting
PDO girl
XKR give-n-ing

OOT glass
PXW GMRS
MKV go-ing(to)
FAU good
CWC government

KQU GPS
GLQ grade-d-r-ing
AXK great
VJI green
QGU grid(coord)

DBG ground-s-ed
ESY group
PVF guard-ed-ing-s
RTB guide-ance
WOM guide-d-s-ing

VBV gun-s
GLG gunpowder
ACB h
JYP had
WMF halt-ed

XPI Ham/ham
PBO hammer

WDK etd
YEB evac-d-ing-ion
AGO evasion

TYV execute-d-ing
YCN execute-ion-s
PKI exercise
RGQ exhaust-ed-ing
CKN expect-ation-s

ILR expect-ed-ing
UVW expedite-d-ing
VXN explode-d-s
CVS exploit-ed-ing
GHY explosive-s

RFL held
LRI helicopter-s
RXG her
JGX here
JCY hidden

XHM hide
JOE high speed burst
JYM high-er-est
KIG highway-s
GQM hill/mt/rise-s

DYG him
BBF his
UWT hit-ting
CUJ hold-er-ing
EIY hole

ASX home
QXV hook
SDR hospital-s
GAI hostile
AQX hour-s-ly

VVB how
EGB hundred
CJJ i
BMO ice
BRR identification

VSP identify-ed

WSY ford-able
VTO forearm
ONV forest-s

WWU fortify-ication
KCB fortify-ied
GNL forward-ed-ing
RQK found
OSM four

DTJ fracture
LGA frag
YWR freedom
DUA freeze (ing)
OEU frequency modulation

UXE ing
UUT initial-ed-ly
BVN insect
HVY insert-s-ed-ing
TBX install-d-ation

TFB instrumental
FWK intact
KSL intend-ed-tion
SNN intercept-d-or
UJJ intercept-ion-s

BFA interfere-d-ing
BDO interfere-nce
FEM internal
AQB interrogate-d
MGS interrogate-ion

ASL intersection
BWC interval
AFB is
FLH isolate-d-ing
FMX issue-d-ing-s

HLL it
BOM item-s
SXC j
RQM jacket
RYE jam-med-ing

ONU January

SOD hand
CBN handbook
BRF handi-talkie

KPV harass-ed-ing
KGH harass-ment
MVY hate
RVE have/has
MGO he

YSV head-ed-ing
DGP headphones
BPG headquarters
HBF hear-ed
NSG heavy

VRU last
VAQ lateral
UBP launch-d-r-ing
ECB law
KKN lay/laid

HTI lead-er-ing
CHQ leaflet-s
IQB least
DTS leave/left
MUF left(of)

PUV leg
EQD letter
OPC level
BRC liaison
BBC library

SEA life
BIR light-er-ng-d-s
DLE like-s
CIC limit-ed-ing-s
QLP line-d-s(of)

GYU list
MUU listen-ed-ing
EQX litter-s
DRS load-s-ed
EKQ local-ize-d

TMF locate-d-ing-s

AQN IDY
BEJ ied
DPX ies
NDR if

WJC ill-ness
FUN immediately
UGQ imminent
NJL immobile-ize-d-ing
OGG impact
AXI impassable

PWR impossible
NFY improve-d-ing
LET improve-ment
PNX improvised
NGB in

LLD in-accordance-with
FKG inadequate
MMS inch-es
MIQ incident-al-s
ING increase-d-g-s

LWV indigenous
WQI indirect
INF individual
BLT infantry
ECJ infiltrate-d-r-s

WPE infiltrate-g-on
JMD inflict
NDU inform-d-ation
PWT infrared
FLX map-ped-ping-s

HPN March/march
XXK market
NTW MARS
SUB match-d-s-ing
MSA max(range)

LUB May
SBL me
MOI meal
WBL means
RQJ measure-d

EOM jeep
SMH job
DQR join-ed-ing
VEY joint

OJQ July
NMO jump-ed-g(off)
KGE junction-s(road)
RYX June
LPI jury
BNC just

GAK k
JFL keep-s-ing
BGQ kerosene
AYQ key
IER kidney

CUD kill
WIO killed in act
LMB kilometer-s
LJG kit
VTA knife

LXJ knot
VIF know-n-ing
ODR knowledge
EJE l
WAB labor-ed-ing

CDM lacerate
XOA lamp
RIT land-ed-ing
SUD large
DOX MRE

HCH multiple
TYE music
BNX my
MKS my location is
OXR n

CMU nail
MSS narrow
VHG near
JAV need-ed-ing
OXM net-ted-ting

GNY locate-ion(of)
DDA lock
NBT log
SVF logistic-al-s

KOU look(for)
YII loss-es-lost
YEL loudspeaker
ULN love
FGJ low-er-est
OLM lower side band

YFC lunch
SWI ly
TRI m
XJA machinegun-s
RTR made

KIM magnetic
EYW magnify-ing
ENF main
GTO main body
CIH main supply rt

YVS maintain-ed
DHG maintain-ing-s
MMU maintenance
LOB make-g/made
DIF male

VOE man-ned-ning
CNL maneuver-ng-s
IYE manual
VSV many
TCI or

MUB orange
VLW order-ed-ing-s
CBI ordnance
TCH organize-ation
UWC organize-d

VXC orient-g-ation
CWW origin-al-ate
FSX other
PCC our
REA out(of)

BQJ measure-ment
EFN meat
BIU mechanic-al
UIW mechanize-d
BKE medical

GKS medium
GBY medivac needed
AST meet (me)(at)
BXG meet-ing
EHH message-s

VHK messenger
MUG meter-s
QTD MFSK
COF microphone
DDL might

KKY mile-s
TCR milk
XJG mine-d(field)
MRB minim-ize-d
NRH minimum

CBT minor
FOD minute-s
RAB miss-ed-ing
XUI missile-s
JGD missing in act

SOM mission-s
KQI mobile-ity
EWN modem
NYV modulation
QWX Monday

ANT money
TRC moon
HLF morale
MVR more
EUX more

KTS morning
QGO Morse
MCX most
CBK mother

CDS network
JMI neutral-ize
EPN neutralize-d-ng
VLA new
PVK news

SPX newspaper
BXW night-s
CXV night-vision
NUW nine
ETM no fire line

LOG no/non/neg
BSU none
AXL north-ern(of)
NKG not
WFQ not later than

PVI note
KOW notebook
SCH nothing
QHO notify-ied-ies
IQD November

HRY now
UFJ nuclear
CPK number-s
QDA o
MSG objective-s

QUR observation(post)
SDF observe-d-r
TYT observe-s-ing
CQJ obstacle-s
MFW occupy-ation

GYD occupy-ed-ing
VUP October
VCD odor
IEK of
PNS offense-ive

DEV officer-s
ALS oil
WJX on
SRA one

TKS outpost
XHR over-age
KWH over-run
OEO overlay
MTO own

HKG p
XYN pacific
TGK packet
XGM PACTOR
KQW panel-s

XDN pants
OAM paratroop-s
MOQ part-s
SWP party-ies
LUH pass-able

ABT pass-age-ing
SSC pass-es-ed
MFJ password
QSG past
UDV patrol-ed-ing-s

XVC PBBS
YEM pen
BHN pencile
MQV penetrate-d-ing
BDL penetrate-ion

TIH people
TQJ pepper
SDN per
ADM percent-age
HBU perimeter

QOF period
TNU permission
VOM permit-ed-ing
XIA PGP
OVB phase

DES photograph-ed-er-ing
LES pick
NLI pick up
MGD piece-d-ing-s

NIC motor-ized
YCT mount-ed-ing
DKT mountain-ous
DHO move-d-ing
XSR move-ment
NWR point-s-ed-ing

FCA police
RLR port(harbor)
LKP position-s-ed
GVU postpone-d-s
OMW postpone-ing

LSR possible
LFK post office
DJL post-s
IGE pound-s
EHO powder

KOH preparation
FJA prepare-d-ing-s
UFX prevent
QVH priest
BRW primary

ERP print
CUI printer
NPJ priority
XBJ privacy
VVW private

VXQ prize-s of war
DNO probable
EWS process-ed-ing
EMA professional
ANQ progress-ed-ing

SNL propane
RFH protect-d-s-ing
NIY protect-ed-ion
AVG protect-ive
VHA provide-d-ing

YOP provide-sion
BND province
JXK PSK

JJQ only

NCL open-ed-ing
AEC operate-ion-s
UHV operate-s-al
OYD opportunity-s
JGS railroad-s

QDS rain
MKW range
EBV rate
NME ration-s-ed
DSP reach-ed-s-ing

VQO read
HWQ ready
FAC rear
PIQ receive-r-d-ng
HPH recognize-d-ing

DAI recon
YNE record
YCV recorder
JAS recover-ed-y
RKJ red

VXP reduce-g-tion
VJB reference (to)
BOE reference-d-s
AOD refuge
EEK refugee(s)

NJP regiment-ed-s
XTO registration
ODW regroup-ed-ng
GIG reinf-ed-s-ing
CTP reinf-ment-s

VQP relay-ed-ing
IDA release-d-ing-s
DUN relief/relieve
TDV remain-ing
ECC rendezvous

XIG reorganize-d-ation-s
VLJ repair-ed-ing
YWL repeat-ed-ing

QMC pipe line

VCY pistol
YAR place-d-ing
RCT plan-s-ed-ing
ARR platoons
EGP ridge-d-s

EDV rifle-s
UVG right(of)
MGJ riot-s-ed
FSW river-stream-s
RWE road-s

AKG roadblock
DNP round-s
VRY route-s-d-ing
SXM RTTY
VWO rucksack

UIT rug
BOX ruin-s-ed
CUQ s
BPO sad
FIR safe-r-ty

YFG said
RAD salt
UPW Saturday
WVN saw
DEO schedule-d-ing-s

GJL school
LCM scope
TCB scout-ed-ing-s
PDD screen-d-ing-s
HBJ search-d-ng-s

YFU second-s-ary
ODI secret
HLI section-s
NJS sector-s
DJS secure-d-ing

QAB secure-ity-s
ERF send-ing/sent
DAP separate-d-ion

DEU psych
SJP pure

UTL purification
GWD pursue-d-ing-t
EDQ put
HYQ q
FUS quell

QDO quest
QFY question-ed
VVD quiet
EUL quit
GLD quiz

FBF r
LHJ rad(rad/hr)
RLI radar
VVP radiation

DRG radiator
SJD radio-s
OOC radiological
FXN raid-er-ing-s
BLC railhead

DQX sister
FNG sit/sitrep
KUW site-s
VSE six
DGS sleep-ing

DSM slope-s-ing-d
WEN slow-ed-ing
VAI small-er-est
DQJ smoke-d-ing-s
TDX sniper

SAO snow
YMD so
LQV socks
QOU solar
BJO solder

DTX soldier
YXT some
ISG soon-er-est

HBR repeater (radio)
AWQ repel-ed-ing-s

CWK replace-d-ing
OBT replace-ment
RKE report-ed-ing
BCX report-s(to)
LHR repulse-d-ing

WOW request-d-g-s
CIQ require-d-ing
KDF require-ment
QLU rescue
JOH reserve-d-g-s

NDF resist-ed-ance
CBL resist-ing
WNF rest
AAU restore-s-d

FJB restrict-ion
DKL restrict-ive
XED result-s-ed-ng
FPD resupply-d-s-g
YJT return-ed-ing

BBJ survey-ed-s
UQJ survival
IEE suspect-d-ing
CCX swallow-ed(s)
OWN switch-ed-ing

UGL swollen
BWR symmetric
BSK t
AAJ tablet
EAH tactic-s-al

IJV take-ing/took
GBF talk
QVD tank destroyer-s
MTS tank-er-s
QQY tape

BWP target-s
PWX task(force)
FEJ taxi

EEN September
GIX sequence

LXG service-d-able
XSC seven
VBE shall
IGD shell-s-ing-ed
GAR shelter

DEX shirt
OBY shoot-ing/shot
AQP short-ed
LJX shortwave
UAV shoulder

AYO sick
OIA sieze-d-ing
ERS sieze-ure-s
BRN sight-ed-ing

XSD sign-ed-ing
NQF signal-ed-ing
OPJ silent-ly-ce
OCM simplex
UFL simulate-d-ing

PIF tomorrow
UMX ton-s-age
CVO tone
GTP tonight
NBX tooth

BKC top secret
VAG total
GDX tow-ed-ing
EJY toward
VNQ town/village-s

ESL trace-r
JWD traffic
EJN trail-er-d-ing-s
EXU train-ed-ing-s
SSM track-ed-ing

SGE transmission
TFF transmit-ed
UTT transport-ed

DMT sorry

DBQ sortie-s
TRH south-ern(of)
UTV speak
VXR spearhead
GGV spice

WSP spine
JRF sporadic
PVJ spot-ted
NHG squad-s
THT stand-ing(by)

CXW start-ed-ing
SBJ station-ed-s
CNP status(report)
EBC stay-ed-ing
BRG stole-n

DPV stop-ped-ing
RDI storm
IKN storm-ing-(s)
RLY street
PGD strength-en

AEO strike/struck
IDR strong-er-est
GPO submit-ed-ing
FUE success-ful-ly
VQS such
DDP Sunday
XCX sunrise
NLY sunset
RPX superior-ity
VFI supplement-ed-ary

HQN supply-d-s-g
QGK support-s-d-ing
ERG surface-d-ing
KQF surgery
ITL surround-ed-ing

RMS surveillance
NMJ violet
XUL virus
FNO visible-ity

VHX team-s-ed

KBB tear-gas
DCG ted
QBF teen
VBN teeth
QPH telephone-s

UQS television
KOF temple
ICB tentative
IWX terminate-d-g
NNQ terrain

TGR territory
ALU terrorism
YIF terrorists
WJN that
WEU the

NPM their
SKW them
FYT then
DTP there
WHI these

RST they
DVY this
XMX those
THP thousand
KUN three
XXF THROB
DMA through
BEG thunder
VPT Thursday
BQH tide

RTT time
PWA time-d-ing
CXL tion
XSI tire
SDA to(be)

NWC today
OSA yet
KMR you-r
GRV z

FFN transport-g-s

TEA transportation
NPD trapped
NKS trouble
NNX truck-ed-s-ing
ILT Tuesday

EXV turn-ed-ing
PLG two
MOA u
NIP unable
VBP unarmed

TXA unauthorized
XHD under
NVB undercover
QPU understand
UNF understood

NAW unharmed
PPQ unidentified
MOP unite-d-s-ing
BGR until(further notice
SEJ unusual

LYS up
QXP upper side band
IPT us
YYT use-d-g
NQK UTC
OVQ V
HUG vaccine
RMT valley
FES vegetable
CKT vehicle-s

MCJ verification
LKE verify-ied
MAH vertical
NQG very
VOX video

TRD violate-s-d

LIJ volt-s

ITA voltage
KBW vulnerable
GNK w
WFW wagon
FEU wait-ed-ing

HJE war
FIK warehouse
HHM was/were
YVH water
XGJ watts

NID wax
GOE waypoint
PMC we/us
UOD weak-en-ness
ANK weapon-s

HLQ weather
CYE Wednesday
EHY week-s
ARS weekend
OLG well

HBW west-ern(of)
EGV what
VVQ wheel-ed
HHW when
HRD where
IIC which
GYV white
TIX who
YYW wife

BRA wild
LGO wilderness
NKE will
ENM wind
RRN WinLink

JTF wire-s-d-ing
IQI with
OTU withdraw-ing-s
GEY wood-s-ed
EHF work-s-ed-ing

YNQ zap

WYO zero
ALX zip
WQX zipper
XLC zone-s-ing
AHG zoo

WOD wound-ed-s
RYA wreck-ed-s
NES wrist
WYU wrong
ADK x

XJC x-mit
BWJ x-ray
CFU y
MCB yard-s
WHR yellow
FPR yes
IMY yesterday

DECODE LIST:

AAJ tablet
AAR fruit
AAS as
AAU restore-s-d
ABO et

ABT pass-age-ing
ACB h
ACT conserve-d
ADD bite
ADK x

ADM percent-age
ADY April
AEC operate-ion-s

AEO strike/struck
AEX clear-ed-ance

AFB is
AFK abort-s-ed-ing
AFL c
AGD estimate-ion
AGO evasion

AHG zoo
AHW fade-ed-ing
AJQ altitude
AJS am in pos (to)
AKG roadblock

AKO collect-ed-ion
AKP far
ALS oil
ALU terrorism
ALX zip

AME barrage-s B
ANG foe
ANK weapon-s
ANQ progress-ed-ing
ANT money

AOD refuge

AWQ repel-ed-ing-s
AXI impassable
AXK great
AXL north-ern(of)
AYA brother

AYH extend-extent
AYO sick
AYQ key
BBC library
BBF his

BBJ survey-ed-s
BCX report-s(to)
BCY fog

BDB battalion-s
BDI embrace-d-ing

BDL penetrate-ion
BDO interfere-nce
BEG thunder
BEJ ied
BFA interfere-d-ing

BFF airdrop-s
BFH crane
BFR direct-d-on-ng
BGQ kerosene
BGR until(further notice

BHN pencil
BIR light-er-ng-d-s
BIU mechanic-al
BJO solder
BKC top secret

KE medical
BKF approach-ed-s
BLC railhead
BLT infantry
BLY city

BMA clearing-s

BQJ measure-ment
BQL fight-er-ing
BRA wild
BRC liaison
BRF handi-talkie

BRG stole-n
BRN sight-ed-ing
BRR identification
BRW primary
BSK t

BSU none
BSV bike
BTB break/broken

BTJ anti-
BUJ cancel-led-s

BUK dangerous
BVF boobytrap
BVN insect
BWC interval
BWD er

BWJ x-ray
BWP target-s
BWR symmetric
BXG meet-ing
BXW night-s

CAU designate-ion
CBI ordnance
CBK mother
CBL resist-ing
CBN handbook

CBR car
CBT minor
CCD administrative
CCI assist-ance
CCX swallow-ed(s)

CCY attach-ed-ing

APK arm-s-ed
APN cost
AQB interrogate-d
AQN IDY

AQP short-ed
AQX hour-s-ly
ARH block-ed-ing
ARR platoons
ARS weekend

ARV enlist-ed-ing
ASH confidential
ASL intersection
AST meet (me)(at)
ASX home

ATR battery-ies
AUQ delay-ed-ing-s
AVF accomplish-ed
AVG protect-ive
AVL border

CKT vehicle-s
CMG cup
CMU nail
CND gate
CNK drive

CNL maneuver-ng-s
CNP status(report)
COF microphone
CPC commit-ment

CPK number-s
CPL eight
CQJ obstacle-s
CQQ depot
CSH administration

CTP reinf-ment-s
CTW counterattack-ed-s
CTX connect-ion
CUA extend-d-g-ive
CUD kill

CUI printer

BMO ice
BNC just
BND province
BNJ approve-al-d-ing

BNP bad
BNX my
BOC decontaminate-d-ion
BOE reference-d-s
BOM item-s

BOX ruin-s-ed
BPG headquarters
BPI final
BPJ escape-d-ing
BPO sad

BPP ambulance
BPS aircraft
BPV detach-ment-s
BPX ask
BQH tide

DJL post-s
DJS secure-d-ing
DKL restrict-ive
DKT mountain-ous
DKU fail-ed-ing

DLE like-s
DMA through
DMR aviator/pilot
DMT sorry

DNO probable
DNP round-s
DOX MRE
DPV stop-ped-ing
DPX ies

DQJ smoke-d-ing-s
DQR join-ed-ing
DQX sister
DRG radiator
DRR deploy-ed-ing

DRS load-s-ed

CDA by
CDL crossroad-s
CDM lacerate
CDR capture-d-ing

CDS network
CEQ able/ability
CEV decrease-d-ing
CFQ alley
CFU y

CGI blister
CHQ leaflet-s
CIC limit-ed-ing-s
CIH main supply rt
CIL arrive-al-d-ing

CIQ require-d-ing
CJA atlantic
CJJ i
CJL company
CKN expect-ation-s

EIL check-d-ing-s
EIY hole
EJE l
EJN trail-er-d-ing-s
EJY toward

EKQ local-ize-d
EKY computer
EMA professional
EMI adjacent

ENF main
ENM wind
ENO connect-ed-ing
ENU alternate
EOF an

EOM jeep
EPN neutralize-d-ng
EQD letter
EQG corporation
EQX litter-s

ERF send-ing/sent

CUJ hold-er-ing
CUQ s
CUW dispose-d-al
CVO tone

CVS exploit-ed-ing
CVV deadline-d-ing
CWC government
CWK replace-d-ing
CWW origin-al-ate

CXL tion
CXV night-vision
CXW start-ed-ing
CYE Wednesday
DAI recon

DAP separate-d-ion
DAW burst-ing-s
DBG ground-s-ed
DBO add-ing-s
DBP enemy

DBQ sortie-s
DCG ted
DDA lock
DDL might
DDP Sunday

DEO schedule-d-ing-s
DES photograph-ed-er-ing
DEU psych
DEV officer-s
DEX shirt

DGP headphones
DGS sleep-ing
DHG maintain-ing-s
DHO move-d-ing
DIF male

DJE b
FJB restrict-ion
FKG inadequate
FLB change to/from ciphe
FLH isolate-d-ing

DSM slope-s-ing-d
DSP reach-ed-s-ing
DTJ fracture
DTP there

DTS leave/left
DTX soldier
DUA freeze (ing)
DUJ FEMA
DUM authorization

DUN relief/relieve
DVX convoy-s
DVY this
DYG him
EAH tactic-s-al

EBC stay-ed-ing
EBV rate
ECB law
ECC rendezvous
ECJ infiltrate-d-r-s

EDC ditch
EDQ put
EDT boots
EDV rifle-s
EDW complete-d-ing

EEK refugee(s)
EEN September
EFN meat
EGB hundred
EGP ridge-d-s

EGV what
EHF work-s-ed-ing
EHH message-s
EHO powder
EHY week-s

EII combat
GQK compound
GQM hill/mt/rise-s
GQN culvert
GRV z

ERG surface-d-ing
ERM frequency-ies
ERP print
ERS sieze-ure-s

ESL trace-r
ESY group
ETL FRS
ETM no fire line
EUL quit

EUX more
EVE emergency
EWN modem
EWS procees-ed-ing
EXU train-ed-ing-s

EXV turn-ed-ing
EYW magnify-ing
FAC rear
FAU good
FBF r

FBN barometer
FBP drug-s
FCA police
FDV boundary-ies
FEJ taxi

FEM internal
FEQ encounter-d-ing
FES vegetable
FEU wait-ed-ing
FFI for

FFN transport-g-s
FGJ low-er-est
FHG embank-ment
FIK warehouse
FIR safe-r-ty

FJA prepare-d-ing-s
IFU dig/dug (in)
IFW bed
IGD shell-s-ing-ed
IGE pound-s

FLX map-ped-ping-s
FMX issue-d-ing-s
FNG sit/sitrep
FNO visible-ity
FNU fire-d-ing-s

FOD minute-s
FPD resupply-d-s-g
FPN church
FPR yes
FRE bulldozer

FRS bit
FSW river-stream-s
FSX other
FUE success-ful-ly
FUN immediately

FUS quell
FWJ conservation
FWK intact
FXN raid-er-ing-s
FYT then

GAI hostile
GAK k
GAR shelter
GBF talk
GBT deliver-ed-ing

GBY medivac needed
GDM chance
GDX tow-ed-ing
GEY wood-s-ed
GGV spice

GGW disregard-d-ing
GHD blue
GHK craft
GHY explosive-s
GIG reinf-ed-s-ing

GIX sequence
GJL school
GKS medium
GLD quiz
GLG gunpowder

GTO main body
GTP tonight
GVA admit
GVU pospone-d-s
GWA come-ing

GWD pursue-d-ing-t
GXU between
GYD occupy-ed-ing
GYU list
GYV white

HBF hear-ed
HBJ search-d-ng-s
HBR repeater (radio)
HBU perimeter
HBW west-ern(of)

HCH multiple
HDF effect-ive-s
HFM en
HHC bivouac-ed-ing
HHM was/were

HHW when
HHX counterattack-ing
HJE war
HKG p
HLF morale

HLI section-s
HLL it
HLQ weather
HLV agent-s
HME at (once)

HPH recognize-d-ing
HPN March/march
HQN supply-d-s-g
HRD where
HRY now

HSN arrange-ment
HTI lead-er-ing
HTP amputate
HUG vaccine
HVY insert-s-ed-ing

IGQ fuse-d
IIB engine
IIC which
IJV take-ing/took
IJW defense-ive

IKN storm-ing-s
ILR expect-ed-ing
ILT Tuesday
IMY yesterday
INF individual

ING increase-d-g-s
IPF capacity-ies
IPT us
IQB least
IQD November

IQI with
IQN attach-ment
ISG soon-er-est
ISL distance-s
ITA voltage

ITD continue-d-ing
ITL surround-ed-ing
IWG cut-off
IWX terminate-d-g
IXN critical-ly

IYE manual
JAS recover-ed-y
JAV need-ed-ing
JBI airborne DF platform
JBY favor-s-ed

JCY hidden
JFL keep-s-ing
JGD missing in act
JGS railroad-s
JGX here

JIA capsule
JJK chemical-s
JJQ only
JKC August
JKE CB radio

GLQ grade-d-r-ing
GML five
GMS friend-ly
GNK w
GNL forward-ed-ing

GNY locate-ion(of)
GOE waypoint
GOU arrow
GPF back-ed-ing
GPO submit-ed-ing

GPV cease-s-d-ing
JTF wire-s-d-ing
JVM before
JWD traffic
JWS all

JXK PSK
JYE f
JYM high-er-est
JYP had
KBB tear-gas
KBW vulnerable
KCB fortify-ied
KCQ elevate-d-ion
KCU command-er-ed-ing-s
KDF require-ment

KDQ destination
KEM collapse
KFE advance-d-ing
KGD alt supply pt
KGE junction-s(road)

KGH harass-ment
KGR band
KHV call
KIE drip
KIG highway-s

KIM magnetic
KKG Friday
KKN lay/laid
KKO cook
KKY mile-s

HWQ ready
HYQ q
IAE bomb-d-r-ng-s
ICB tentative
ICX gasoline

IDA release-d-ing-s
IDR strong-er-est
IEE suspect-d-ing
IEK of
IER kidney

IEU agree
LDV engineer-d-ing-s
LEP detach-ed-es
LES pick
LET improve-ment

LFC elk
LFJ general-ize
LFK post office
LGA frag
LGO wilderness
LHJ rad(rad/hr)
LHR repulse-d-ing
LID assault
LIH a
LII ammunition

LIJ volt-s
LJG kit
LJK district-s
LJX shortwave
LKE verify-ied

LKP position-s-ed
LLD in-accordance-with
LMB kilometer-s
LNA coast-s-al
LNJ and

LNN could
LOB make-g/made
LOG no/non/neg
LPI jury
LQV socks

JKO comma
JKT defend-ed-g-s
JLY class-ify
JMD inflict
JMI neutral-ize

JML asap
JNU camouflage-d
JOE high speed burst
JOH reserve-d-g-s
JRF sporadic

JTA concentrate-s-ion
MGD piece-d-ing-s
MGJ riot-s-ed
MGO he
MGS interrogate-ion

MHN beer
MHQ fuel
MIQ incident-al-s
MKS my location is
MKV go-ing(to)
MKW range
MMS inch-es
MMU maintenance
MNM future
MOA u

MOI meal
MOP unit-ed-s-ing
MOQ part-s
MOS fold-ing
MQE clean-ed-ing

MQV penetrate-d-ing
MRB minim-ize-d
MRS force-s-d-ing
MSA max(range)
MSG objective-s

MSS narrow
MTN determine
MTO own
MTS tank-er-s
MTV from

KMB damage-d-ing
KMR you-r
KOF temple
KOH preparation
KOO consolidate-d

KOU look(for)
KOW notebook
KPM CW
KPV harass-ed-ing
KQF surgery

KQI mobile-ity
KQU GPS
KQW panel-s
KSJ feet/foot
KSL intend-ed-tion

KTD caliber
KTS morning
KUN three
KUW site-s
KWH over-run

KYH accomplishing
LAD divert-ed-ion
LAG communicate-d-ions-s
LAO deer
LCM scope

LDH boy
NID wax
NIP unable
NIY protect-ed-ion
NJL immobile-ize-d-ing

NJP regiment-ed-s
NJS sector-s
NKD contaminate-ion-ing
NKE will
NKG not

NKS trouble
NKX column-s
NLF baud
NLI pick up

LRB electric-al-ity
LRI helicopter-s
LSR possible
LSY abdomen
LUB May

LUH pass-able
LWB e-mail
LWF approach-ing
LWV indigenous
LWY cross-ed-ing

LXG service-d-able
LXJ knot
LYS up
MAH vertical
MAV authorize-d-ng

MAY acknowledge-d-ment
MCB yard-s
MCJ verification
MCX most
MDA cache'

MDT canal-s
MDV DTMF
MEQ contact-ing
MFJ password
MFL deception

MFW occupy-ation
OFX contaminate-d
OGG impact
OHB coord-s-d-ing
OIA sieze-d-ing

OJQ July
OLG well
OLM lower side band
OLV direction finding
OMU facility

OMW pospone-ing
ONA accident-al-s
ONT diesel
ONU January

MUB orange
MUF left(of)
MUG meter-s
MUU listen-ed-ing
MVR more

MVY hate
MXI beach-ed-s-ing
MYY east-ern (of)
NAW unharmed
NBT log

NBX tooth
NCF downed acft
NCL open-ed-ing
NCM continue-ation
NDF resist-ed-ance

NDM equip-ing-s
NDR if
NDU inform-d-ation
NES wrist
NFC dog

NFY improve-d-ing
NGB in
NHG squad-s
NHI control-d-ing-s
NHT commit-ted-ing

NIC motor-ized
PNS offense-ive
PNX improvised
PPQ unidentified
PQA article

PQJ disperse-d-g-l
PRA envelope
PSA add-ed-itional
PTK breast
PUV leg

PVC act-ive-ivity-s
PVF guard-ed-ing-s
PVI note
PVJ spot-ted

NLJ cloth-es-ing

NLN father
NLY sunset
NME ration-s-ed
NMJ violet
NML gas-ing-ed

NMO jump-ed-g(off)
NNQ terrain
NNX truck-ed-s-ing
NOH destroy-d-ing-s
NON child/children

NPD trapped
NPJ priority
NPM their
NQF signal-ed-ing
NQG very

NQK UTC
NQM flesh
NRH minimum
NRY d
NSG heavy

NSM airmobile
NTW MARS
NUR automatic
NUW nine
NVB undercover

NWC today
NWR point-s-ed-ing
NYL during
NYT extract-s-d-ing
NYV modulation

OAM paratroop-s
OBT replace-ment
OBY shoot-ing/shot
OCM simplex
ODI secret

ODR knowledge
ODW regroup-ed-ng
OEO overlay

ONV forest-s

OOC radiological
OOT glass
OPC level
OPJ silent-ly-ce
OPK bridge-d-ing-s

OSA yet
OSM four
OTF extend-sion
OTL blow-ing (up)
OTU withdraw-ing-s

OVB phase
OVQ V
OWI book
OWN switch-ed-ing
OXM net-ted-ting

OXR n
OYD opportunity-s
OYT flank-ed-ing-s
OYY crystal
PAO field-s

PBO hammer
PCC our
PCQ alert-ed-ing
PDD screen-d-ing-s
PDO girl

PDU breakfast
PDV encrypt-ed-tion
PGB army-ies
PGD strength-en
PHF depart-d-g-ure

PIF tomorrow
PIN expose-d-ing
PIQ receive-r-d-ng
PJK dinner
PKE contact-ed-s

PKI exercise
PKN complete-ion
PLG two

PVK news

PWA time-d-ing
PWE employ-ed-ing
PWR impossible
PWT infrared
PWX task(force)

PXF death
PXW GMRS
PYE charge-d-s
PYN ed
QAB secure-ity-s

QAH ache
QAO es
QBF teen
QDA o
QDO quest

QDS rain
QEP area-s
QEY end-ed-ing
QFY question-ed
QGK support-s-d-ing

QGO Morse
QGU grid(coord)
QHB antenna
QHO notify-ied-ies
QKQ deny-ied-ies

QKT flare-d-s
QLP line-d-s(of)
QLU rescue
QMC pipe line
QMD at

QMS channel
QNN attack-ed-ng-s
QNS available
QNT arrange-d-ing
QOF period

QOU solar
QOW disable-d-ing
QPH telephone-s

OEU frequency modulation
OFP carry-ied-iers

OFT female
QRC gallon-s
QSF divide
QSG past
QTB December

QTD MFSK
QTF deploy-ment-s
QTQ checkpoint
QUR observation(post)
QVD tank destroyer-s

QVH priest
QWX Monday
QXH concentrate-d-ing
QXP upper side band
QXV hook

QYA candle
RAB miss-ed-ing
RAD salt
RBP conduct-ed-ing
RBR air (fld/port)

RBT bug(s)
RCL do
RCT plan-s-ed-ing
RDI storm
REA out(of)

RFH protect-d-s-ing
RFL held
RGL apply
RGQ exhaust-ed-ing
RIF flexible-ity

RIL bone
RIT land-ed-ing
RKE report-ed-ing
RKF favor-able-ing
RKJ red

RLI radar
RLR port(harbor)

PMC we/us
PMT approximate-ly

PMV get-ting
RUS age
RVE have/has
RVX day-s/date-s
RWA battery charger

RWE road-s
RWH challenge-d-ing
RWI binoculars
RXG her
RYA wreck-ed-s

RYE jam-med-ing
RYX June
SAO snow
SAP capable-itiy
SBE disapprove-d-al

SBJ station-ed-s
SBL me
SBO belt
SCH nothing
SDA to(be)

SDF observe-d-r
SDN per
SDR hospital-s
SEA life
SEF front-age-ally

SEJ unusual
SGE transmission
SIX blow-n
SJD radio-s
SJP pure

SKW them
SMH job
SMN displace-d-ing
SNL propane
SNN intercept-d-or

SOD hand
SOF envelop-ed-ing

QPU understand
QPW eliminate-d-ing-s

QQY tape
TBO axe
TBX install-d-ation
TCB scout-ed-ing-s
TCH organize-ation

TCI or
TCR milk
TDV remain-ing
TDX sniper
TEA transportation

TFB instrumental
TFF transmit-ed
TGK packet
TGR territory
THP thousand

THT stand-ing(by)
THX bow
TIH people
TIX who
TKS outpost

TKW classification
TLH box
TMF locate-d-ing-s
TNU permission
TOF CTCSS

TON battle-s
TQJ pepper
TRC moon
TRD violate-s-d
TRH south-ern(of)

TRI m
TRM alarm
TSG eta your loc
TTM consolidation
TUQ base-d-s-ing

TXA unauthorized
TYE music

RLY street
RMS surveillance
RMT valley

RNI desk
RNT flame-d-ing-s
RPU begin
RPX superior-ity
RQJ measure-d

RQK found
RQM jacket
RQP change
RRN WinLink
RST they

RSU (SPACE / NULL)
RTB guide-ance
RTP civilian-s
RTR made
RTT time

RUN down-ed
UHM cover-ed-ing
UHV operate-s-al
UIT rug
UIW mechanize-d

UJJ intercept-ion-s
UKM fish
UKR azimuth (of)
ULN love
UMF bleed

UMX ton-s-age
UNF understood
UNU circle
UOD weak-en-ness
UPQ fatal

UPW Saturday
UQJ survival
UQS television
UTF flight
UTL purification

UTT transport-ed

SOM mission-s
SPG FeldHell
SPR demonstrate-d-s-ion

SPX newspaper
SQG assist-ed-ing
SRA one
SSC pass-es-ed
SSM trak-ed-ing

SUB match-d-s-ing
SUD large
SVF logistic-al-s
SVI fallout
SVW division-s

SWI ly
SWP party-ies
SXC j
SXM RTTY
TAB collect-ing-s

TAT close-d-ing
VIF know-n-ing
VJB reference (to)
VJI green
VJU alive

VKC compromise-d-ing
VKL danger
VLA new
VLJ repair-ed-ing
VLU alternate commo

VLW order-ed-ing-s
VNB be-en-ing
VNQ town/village-s
VNS bank
VOE man-ned-ning

VOM permit-ed-ing
VOX video
VPN cave
VPT Thursday
VQO read

VQP relay-ed-ing

TYT observe-s-ing
TYV execute-d-ing
UAR assemble-d-ing

UAV shoulder
UBL barrier
UBP launch-d-r-ing
UCB emplace-ment
UCK below

UCV camera
UDV patrol-ed-ing-s
UFJ nuclear
UFL simulate-d-ing
UFW conversation

UFX prevent
UGC corps
UGJ drill
UGL swollen
UGQ imminent

UHG extra
WGE error
WHI these
WHR yellow
WIO killed in act

WJC ill-ness
WJN that
WJX on
WLY consolidate-ng
WMF halt-ed

WNC above
WNF rest
WNW displace-ment
WOD wound-ed-s
WOM guide-d-s-ing

WOW request-d-g-s
WPE infiltrate-g-on
WQI indirect
WQX zipper
WRM ALE

WRT but

UTV speak
UUL element-s
UUT initial-ed-ly
UVC bunker-s

UVG right(of)
UVJ ambush-ed-ing
UVO estimate-d-ing
UVW expedite-d-ing
UWC organize-d

UWT hit-ting
UWU correct-ed-ing
UXE ing
VAG total
VAI small-er-est

VAQ lateral
VBE shall
VBN teeth
VBP unarmed
VBV gun-s

VCD odor
VCY pistol
VDJ darkness
VDW can
VEA barbed wire

VEQ amateur
VEV February
VEY joint
VFI supplement-ed-ary
VGV condition-d-ing

VHA provide-d-ing
VHG near
VHK messenger
VHM dump-ed-ing
VHP entrench-ed

VHX team-s-ed
XMX those
XOA lamp
XPI Ham/ham
XQQ follow-ed-ing

VQS such
VRU last
VRY route-s-d-ing
VSD estab-ed contact at(

VSE six
VSP identify-ed
VSV many
VTA knife
VTN ball

VTO forearm
VUP October
VVB how
VVD quiet
VVI amplitude modulation

VVP radiation
VVQ wheel-ed
VVW private
VWE family
VWO rucksack

VXC orient-g-ation
VXN explode-d-s
VXP reduce-g-tion
VXQ prize-s of war
VXR spearhead

VXT axis (of)
WAB labor-ed-ing
WBL means
WDK etd
WDX drop-ping-ped

WEN slow-ed-ing
WEU the
WFC clarification(of)
WFQ not later than
WFT answer-d-ing-s

WFW wagon
YVS maintain-ed
YWL repeat-ed-ing
YWR freedom
YXG coffee

WSP spine
WSY ford-able
WVE been
WVH degree-s

WVK delete-d-ion
WVN saw
WWU fortify-ication
WXW g
WYC deputy-ies

WYO zero
WYU wrong
WYV generator(set)
WYY about
XAD designate-d

XBG cable-s
XBJ privacy
XCX sunrise
XDN pants
XED result-s-ed-ng

XFY are/is
XGJ watts
XGM PACTOR
XHD under
XHM hide

XHR over-age
XIA PGP
XIG reorganize-d-ation-s
XIV food
XJA machinegun-s

XJC x-mit
XJG mine-d(field)
XKJ biological
XKR give-n-ing
XLC zone-s-ing

XME building-s

XSC seven
XSD sign-ed-ing
XSI tire
XSR move-ment
XSX due

XTG am
XTO registration
XUF after
XUI missile-s
XUL virus

XVC PBBS
XVE e
XVI fled
XVY around
XWF act-ion-ing-s

XXF THROB
XXK market
XYN pacific
XYV camouflage-ing
XYY crushed

YAR place-d-ing
YBS camp
YCN execute-ion-s
YCT mount-ed-ing
YCV recorder

YDL boat-s
YEB evac-d-ing-ion
YED establish-d-ng
YEL loudspeaker
YEM pen

YFC lunch
YFG said
YFU second-s-ary
YGA correct (me)
YIC demolition

YIF terrorists
YII loss-es-lost
YJM bypass-d-ing-s
YJT return-ed-ing
YLI counterpart

YXT some
YYL automobile
YYT use-d-g
YYW wife
YYY beyond

YMD so
YMQ demonstrate-d-ing
YNE record
YNQ zap
YOP provide-sion

YOR casualty-ies
YQC effect-ed-ing
YRX black
YSV head-ed-ing
YVH water

YVP driveway

Made in the USA
Monee, IL
02 July 2023